Nanomaterial-Supported Enzymes

Edited by

Inamuddin[1], Tariq Altalhi[2], Jorddy N. Cruz[3], Mohammad Luqman[4]

[1]Department of Applied Chemistry, Zakir Husain College of Engineering and Technology, Faculty of Engineering and Technology, Aligarh Muslim University, Aligarh-202002, India

[2]Department of Chemistry, College of Science, Taif University, 21944 Taif, Saudi Arabia

[3]Adolpho Ducke Laboratory, Botany Coordination, Museu Paraense Emílio Goeldi, Av. Perimetral, 1900, Terra Firme, Belém PA 66077-830, Brazil

[4]Department of Chemical Engineering, College of Engineering, Taibah University, Yanbu, Saudi Arabia

Published by **Materials Research Forum LLC**
Millersville, PA 17551, USA

Published as part of the book series
Materials Research Foundations
Volume 126 (2022)
ISSN 2471-8890 (Print)
ISSN 2471-8904 (Online)

Print ISBN 978-1-64490-196-0
eBook ISBN 978-1-64490-197-7

Distributed worldwide by

Materials Research Forum LLC
105 Springdale Lane
Millersville, PA 17551
USA
https://www.mrforum.com

Manufactured in the United States of America
10 9 8 7 6 5 4 3 2 1

Table of Contents

Preface

This book focuses on the use of nanomaterials and enzymes to support different applications in the materials, food, pharmaceutical and biotechnology industries. It comprehensively covers topics relevant to understanding the development of nanotechnology-integrated biological systems and their innovative functionality. Among several advantages, the enzyme, after being immobilized, becomes more stable, being more suitable for the use in drug delivery systems, in addition to increasing efficiency in catalysis systems. In this book, its chapters cover recent advances in production and properties of nanoenzymes, immobilization of enzymes in nanomaterials and their applicability in the area of disease diagnosis, environmental clean-up, biosensor manufacturing, drug delivery and vaccine production.

Nanomaterials Supported Enzymes aims to explore in their chapters with discussions on a variety of nanomaterials and enzymes to support various applications in the materials, food, pharmaceutical, and biotechnology industries. Leading researchers and specialists from all over the world contribute to the book chapters. This book is a reference text that can be used by postgraduates, biotechnologists, engineers, scientists, professors and environmentalists working in the field of nanotechnology, nanomaterials, biosensors, diagnostics, food industry, drug delivery, vaccine production, engineering, environmental science, and industries. The work reported in the following 9 chapters are as follows:

Chapter 1 describes the various types of nanomaterials i.e. metal nanoparticles, metal oxide, carbonaceous materials (carbon nanotubes, graphene, and activated carbon), that have been used for the immobilization of the enzyme. So that durability, catalytic activity, leaching of the enzyme, and mechanical steadiness are evaluated for their continual operation.

Chapter 2 discusses the production, properties, and applications of nano-enzymes of carbon, zinc oxide, magnetite, copper, and some noble metals in the food industry.

Chapter 3 highlights the literature on nanozymes; their recent fabrication for enzyme mimicking activity and finally their mechanism of actions to enhance enzymatic activities concerned with the food industries.

Chapter 4 discusses the different immobilization methods, nanosupports for immobilization and their uses in the depollution of aquatic environments, as far as nanomaterials supported enzymes are concerned.

Chapter 5 lays an outline with respect to biosensors that are fabricated with nanomaterials as a transducer, where enzymes acting as a bioreceptor, are immobilized on their surface. In addition, the biosensing mechanisms of the enzyme immobilized nanomaterials, their efficiency, detection limit, and sensitivity, are also discussed.

Chapter 6 highlights the different fascinating applications in therapeutics along with the associated mechanism of nanozymes, which are efficient nanomaterials with enzyme-like appearances.

Chapter 7 elaborates the diversified range of nano-material based enzymes, their synthesis methods, modification strategies, and factors influencing the catalytic activity of these enzymes. Therapeutic applications of nano-material based enzymes and their limitations have also been discussed.

Chapter 8 enumerates the importance of how immobilized enzymes are helpful in different biomedical uses and what kind of enzymes and nanoparticles could be hyphenated to take advantage in health care sectors. Different method of enzymes immobilization will also be discussed in details including both physical methods and chemical methods of loading enzymes on nanoparticles.

Chapter 9 focus on nanonzyme and their use in vaccine production and immunization. The nano carrier-based system facilitates the delivery of vaccine antigens to target cells and increases antigen resistance and immunogenicity.

Highlights

- Provides a comprehensive overview of the nanomaterials supported enzymes

- Production and properties of nanoenzymes

- Provides a broad overview of different spheres of applications

- One-shot reference guide for those wishing to learn more on this.

Materials Research Forum LLC
https://doi.org/10.21741/9781644901977-1

Chapter 1

Recent Advances in Enzyme Immobilization in Nanomaterials

Muhammad Hamza[1], Abdul Qadeer[2], Mabkhoot Alsaiari[3,4], Saleh Alsayari[3], Qudsia Kanwal[1], Abdur Rahim[5*]

[1]Department of Chemistry, University of the Lahore, Lahore

[2]Shanghai Veterinary Research Institute, Chinese Academy of Agricultural Sciences, Key Laboratory of Animal Parasitology of Ministry of Agriculture and Rural Affairs, Shanghai, 200241, China

[3]Empty Quarter research Unit, Department of Chemistry, College of science and art in Sharurah, Najran University, Sharurah, Saudi Arabia

[4]Promising Centre for sensors and electronic devices (PCSED). Advanced materials and Nano Research Centre. Najran University. Najran. 11001. Saudi Arabia

[5]Interdisciplinary Research Centre in Biomedical Materials (IRCBM), COMSATS University Islamabad, Lahore Campus, Pakistan.

* rahimkhan533@gmail.com

Abstract

This chapter described the advancements in the development of nanostructured supported material and enzyme immobilization techniques. The functionalized nanomaterials extremely affect the inherent mechanical properties and provide the highest biocompatibility and specific nano-environment surrounding the enzymes for improving enzymes stability, catalytic performance, and reaction's activities. The enzyme immobilization on nanomaterials considerably enhances the robustness and durability of the enzyme for its frequent applications, which reduces the overall expenses of the bio-catalytic process. There are various types of nanomaterials i.e. metal nanoparticles, metal oxide, carbonaceous materials (carbon nanotubes, graphene, and activated carbon), that have been used for the immobilization of the enzyme. So that durability, catalytic activity, leaching of the enzyme, and mechanical steadiness are evaluated for their continual operation.

Keywords

Enzyme Immobilization, Nanomaterials, Biomimetic, Co-factor, Carbonaceous Nanomaterials

List of abbreviations

2D	:	two-dimensional
AC	:	alternating current
AChE	:	acetylcholinesterase
AFM	:	atomic force microscopy
Al	:	Aluminum
Al_2O_3	:	Aluminum Oxide
ALDC	:	Acetolactate decarboxylase
ATP	:	adenosine triphosphate
AuNPs	:	gold nanoparticles
BSA	:	bovine serum albumine
CALB	:	Candida antartica lipase B
ChO	:	choline oxidase
CLEAs	:	cross-linked enzyme aggregates
CNBr	:	Cyanogen bromide
CNS	:	central nervous system
CNTs	:	carbon nanotubes
-COOH	:	Carboxylic Acid
CPRG	:	β-D-galactopyranoside
CPs	:	Conducting polymers
CRL-microbial	:	Candida rugosa lipase
CS	:	chitosan
CTAB	:	cetyltrimethylammonium bromide
Cu	:	Copper
D	:	drain
DFP	:	diisopropylfluorophosphate
DNA	:	Deoxyribonucleic Acid
EC	:	Enzyme Commission
EC	:	endothelial cell
EDC	:	L-ethyl-3- (dimethyl-aminopropyl) carbodiimide hydrochloride
EIS	:	Electrochemical impedance spectroscopy
EISC	:	electrolyte–insulator–semiconductor
EnFET	:	enzyme field-effect transistor
EnNPs	:	Enzymes Nanoparticles
FAD	:	flavin adenine dinucleotide
Fe_3O_4	:	Magnetite
FET	:	field-effect transistor
FMN	:	flavin mononucleotide
FTIR	:	Fourier transform infrared spectroscopy
FTO	:	fluorine doped tin oxide
G	:	graphene
GA	:	glutaraldehyde
GO	:	graphene oxide

GOD	:	glucose oxidase
G-PFIL	:	G-polyethylenimine- functionalized ionic liquid
H_2O_2	:	hydrogen peroxide
H_2Q	:	hydroquinone
H_2SO_4	:	sulfuric acid
Hg	:	Mercury
HIV	:	human immunodeficiency virus
HMP	:	hydrophobic magnetic particles
HNO_3	:	Nitric Acid
HNTs	:	halloysite nanotubes
HRP	:	Horseradish peroxidase
ISE	:	Ion-selective electrode
ISFET	:	ion-sensitive field-effect transistor
LAPs	:	light-addressable potentiometric sensor
LDH	:	lactate dehydrogenase
LDHNSs	:	LDH nanosheets
LDHs	:	Layered double hydroxides
LED	:	light-emitting diode
Mg	:	Magnesium
MIC	:	minimum inhibitory concentrations
MOFs	:	Metal Organic Frameworks
MOSFET	:	metal oxide semiconductor field-effect transistor
MPA	:	mercaptopropionic acid
MWCNTs	:	multi-walled carbon nanotubes
NADH	:	nicotinamide adenine dinucleotide
$-NH_2$	:	Amino Group
$-NH_3^+$	:	Ammonium ion
NHS	:	N-hydroxysuccinimide
Ni	:	Nickel
NO gas	:	Nitric oxide
O_2	:	Oxygen
OxOx	:	oxalate oxidase
PANI	:	polyaniline
Pb	:	Lead
PBCA	:	polybutylcyanoacrylate
PBQ	:	Parabenzoquinone
PCL	:	Pseudomonas cepacia lipase
PEDOT	:	poly(3,4-ethylenedioxythiophene
PEG	:	poly ethylene glycol
pH	:	potential of hydrogen
PLGA	:	polylactic–polyglycolic acid
POU systems:		Point-Of-Use: Under the sink
ppb	:	Part per billion

PPL-animal	:	porcine pancreas lipase
Ppy	:	polypyrrole
PtNPs	:	Platinum Nanoparticles
PVA	:	polyvinyl alcohol
PyrOx	:	pyruvate oxidase
Rct	:	charge transfer resistance
rGO	:	reduced graphene oxide
RML	:	Rhizomucor miehei lipase
RNA	:	Ribonucleic acid
ROS	:	Reactive oxygen species
RSM	:	response surface methodology
S	:	source
SEM	:	scanning electron microscopy
SiO_2	:	silicon dioxide (Silica)
SOD	:	Superoxide dismutase
SWCNTs	:	single-walled carbon nanotubes
TEM	:	transmission electron microscopy
THF	:	tetrahydrofolate
TiO_2	:	Titanium dioxide
TLL	:	Thermomyces lanuginosus lipase
tPA	:	type plasminogen activator
TPP	:	thiamine pyrophosphate
tRNA	:	Transfer RNA
uPA	:	urokinase-type plasminogen activator
XRD	:	X-ray diffraction
Zn	:	Zinc
ZnO	:	Zinc Oxide
ΔG	:	Gibbs free energy

Contents

1. Enzymes and their uses/ applications/ functions

1.2 Definition of enzyme

Enzymes are chemically composed of protein, works as a catalyst in biological reactions, and are known as biocatalysts [1]. Normally, Catalysts speed up chemical or biological reactions. The substances upon which catalysts works are called known as substrates, after reaction completion these catalyst converts starting molecules into various new molecules which are called products. Every metabolic reaction within a cell requires an enzyme that accelerates reaction sufficient for the support of life [2]. The metabolic mechanism depends on catalysts to catalyze every single step. The science of exploring enzymes is known as enzymology and recently an advanced field of pseudo-enzyme analysis has been developed [3, 4]. Currently reported enzymes catalyze more than 5000 metabolic chemical reactions [5]. Some other catalysts are catalytic RNA molecules which are known as ribozymes. Enzymatic specificity arises due to their specific three-dimensional architecture.

An enzyme, like other catalysts, decreases the activation energy of molecules and speeds up the chemical reaction. A few enzymes' activities are millions of times faster than normal reactions they convert substrate to the product at such high speed. For example, the enzyme orotidine 5'-phosphate decarboxylase allows a reaction in milliseconds rather than millions of years [6, 7]. Enzymes are similar to chemical catalysts in that they speed up processes by lowering the activation energy of molecules. They are not utilized in reactions, and they do not change the chemical reaction's equilibrium. Enzymes and catalysts differ in such a way, enzymes are more specific than catalysts. The activity of enzymes is affected by different types of molecules: activators accelerate the activity of enzymes, and inhibitors

Materials Research Forum LLC
https://doi.org/10.21741/9781644901977-1

cease the activity of the enzyme. Various therapeutic poisons and drugs act as an inhibitor for enzymatic activities.

A few examples of commercially used enzymes are the formation of antibiotics. A few household items utilize enzymes to accelerate reactions: In organic washing powder enzymes disintegration of fats, starch, or protein stain on cloth, and in flash tenderizers enzymes disintegration of larger protein molecules into shorter molecules and makes flash easy to digest.

1.2 History & etymology of enzymes

End of the 17th and initial 18th centuries, saliva and secretions from the stomach causes the digestion of meat [8, 9] and convert meats' starch into sugars like plant extracts were known but how it happens, which chemical reactions take parts, and which mechanisms occurred had not been investigated [10].

Enzymes firstly were discovered in 1833 by the French chemist Anselme Payen that was diastase [11]. After a few years of research into the fermentation process of sugar to alcohol, Louis Pasteur hypothesized that the process was catalyzed by a key force housed in yeast cells known as "ferments," and that this force performed its role within living creatures. Further Louis Pasteur concluded that " the process of fermentation of alcohol is an action related along with the corporation of the yeast cell and life, not related with the putrefaction of the cell or death [12]."

Terminology "Enzymes" first reported by German physiologist Wilhelm Kühne (1837–1900) in 1877, this word originated from the Greek word ἔνζυμον, meaning "leavened" or "in yeast", this terminology explains fermentation technique [2, 13-21]. After a while, the enzyme was used to cite "non-living substance" e.g. pepsin, and the ferment was used to cite "chemical action arises through living organism" [22].

In 1987, the first paper of Eduard Buchner was submitted on "the study of yeast extracts". At the University of Berlin, he conducted a sequence of trial investigations and concluded that the fermentation process of sugar through yeast extracts is carried out even when all the yeast cells are dead in the solution mixture [2, 15, 16, 23-29]. He labeled the enzyme as "zymase" which carried out the sucrose fermentation. [30] According to Buchner's following example, most enzymes are labeled based on the reactions which they conduct: the substrate name combined with the suffix-use (e.g., lactase DNA polymerase) [31].

1.3   Nomenclature

Enzymes are classified into two prime categories depending on either the sequence similarity of amino acids or the activity of enzymes.

1.4   Enzyme activity

According to Eduard Buchner enzymes name, most enzymes are named derived from their catalyzed chemical reaction or substrate, with the suffix -a. [2], for example alcohol dehydrogenase, lactase, and DNA polymerase. The same reaction that is catalyzed by the various types of enzymes is known as isozymes [2].

The enzymes nomenclature has been established by the International Union of Biochemistry and Molecular Biology, the EC ("Enzyme Commission") numbers. The entire enzymes are labeled by "EC" along with four-digit numbers that express the grading of enzymes activity. The first digit usually categorizes the enzyme according to its mechanisms while the rest of the digits express enzymes specificity [15, 16].
The principal-level of categorization as:

- **EC 1**, Oxidoreductases: Oxidation and reduction reaction are catalyzed.
- **EC 2**, Transferases: Functional groups are transferred.
- **EC 3**, Hydrolases: Various bonds hydrolysis is catalyzed.
- **EC 4**, Lyases: Different bond cleave rather than oxidation and hydrolysis.
- **EC 5**, Isomerases: Isomerization is catalyzed within the same molecule.
- **EC 6**, Ligases: Bond two molecules through covalent bonding.

The categorization of these primary levels is further divided by various characteristics e.g. the chemical mechanism, products, and substrate. An enzyme is completely identified by its four digits designations e.g. hexokinase (EC 2.7.1.1) [15].

1.5   Sequence similarity

Enzyme Commission division cannot explain sequence relationship. For example, two ligases catalyzing entirely the same reactions, possessing identical EC numbers have entirely contrasting sequences. Independent of enzymes purpose like different proteins have been categorized according to enzymes sequence similarity into various subgroups [32].

1.6   Chemical structure

Enzymes are naturally spherical proteins with complicated structures. Amino acids make up an enzyme. The arrangements of amino acids within enzymes regulate their catalytic activity in a chemical reaction [33]. Besides the structure of the enzyme, its purpose is predicted rather than its unique enzymatic activity just from the enzyme structure alone [34]. The structure of enzymes denatured when exposed to heat or chemical compositions and this disturbance in the enzyme structure is usually liable for the deprivation of activity of enzymatic [35]. Alteration in the structure of enzymes is generally related to a temperature high to enzymes' normal level.

The structure of enzymes is generally larger as compared to their substrate. Enzymes size ranges from 62 amino-acids, for example, 4-oxalocrotonate tautomerase used a monomer [36], to high up to 2,500 amino-acids within animals fatty acid synthase [37]. Just a microscopic segment of enzymes' shape is contributing to the catalysis process known as the catalytic site of the enzyme [38]. The catalytic site exists near one or multiple binding locations where amino acids align substrate. The binding portion and catalytic portion both formulate enzymatic activity and are collectively known as enzyme active sites [39].

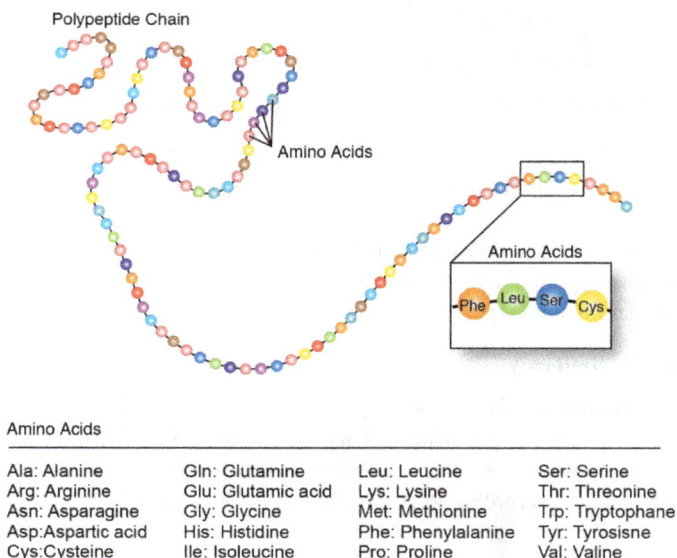

Amino Acids

Ala: Alanine	Gln: Glutamine	Leu: Leucine	Ser: Serine
Arg: Arginine	Glu: Glutamic acid	Lys: Lysine	Thr: Threonine
Asn: Asparagine	Gly: Glycine	Met: Methionine	Trp: Tryptophane
Asp:Aspartic acid	His: Histidine	Phe: Phenylalanine	Tyr: Tyrosisne
Cys:Cysteine	Ile: Isoleucine	Pro: Proline	Val: Valine

Fig. 1: Structure of Enzyme and types of Amino Acids

In the enzymes catalysis process, the amino acid is not involved, in their catalysis process enzyme has an area for binding and aligning catalytic cofactors [39]. The structure of enzymes contains allosteric sites in which the attachment of small molecules results in the change of their conformation which escalates or suppresses enzymatic activity [40].

A few RNA-based biological catalysts are known as ribozymes that react alone or with proteins e.g. protein and catalytic RNA components [41].

1.6.1 Co-factor

Various enzymes do not require supplementary items to express their full enzymatic activity. The remaining enzymes need a non-protein substance known as co-factors that bind with the enzyme for its activity [42]. This substance may be inorganic (e.g., iron-sulfur clusters and metal ions) or organic compound (e.g., heme and flavin). These inorganic and organic co-factors show numerous functions; for example, in inorganic cofactor metals, ion favors in nucleophilic molecule stabilizing on active position of the enzyme [43]. On the other hand, an organic co-factor may be a co-enzyme that is liberated from an active position of enzymes within the reaction, and the organic prosthetic group that is lodged with an enzyme. Generally, the prosthetic group is attached with a covalent bond (e.g., biotin within enzyme as pyruvate carboxylase) [44].

An enzyme-containing cofactor is (e.g. carbonic anhydrase that utilizes zinc atom as a cofactor) attached as a portion of enzymes active site [45]. These attached molecules through covalent bonding are generally present in enzymes active-site and take part in the catalysis process (e.g., Heme and flavin are co-factors usually carried out redox reactions) [16].

The enzymes, which require cofactor and they do not show one bond, are known as "apoproteins or apoenzymes". An enzyme that requires a cofactor along with it for enzymatic activity is known as "haloenzyme or holoenzyme". The term holoenzyme applies to the enzyme, which contains different protein monomers (e.g., DNA polymerases; the holoenzyme has an overall complex structure consisting of overall subunits required for activity) [39].

1.6.2 Co-enzymes

Co-enzymes consist of small organic molecules, which may be covalently bonded or loosely attached to an enzyme. The function of co-enzymes is generally the transportation of chemical groups from enzyme to enzyme (e.g., NADPH, ATP, and NADH) [46]. A few co-enzymes (e.g., FMN, FAD, TPP, THF) are acquired from vitamins. The coenzymes are not produced within the body and these enzymes' homogenous molecules must be obtained from our diet [2, 14-16, 46, 47].

Co-enzymes are a unique category of substrate that chemically changes their structures or arrangements according to enzymes' actions. Co-enzymes are unique groups of molecules, or substrates that are identical to numerous enzymes (e.g., more than 1000 enzymes are used as co-enzyme of NADH) [48, 49].

Co-enzymes are permanently reproduced and their quantity remains at a uniform level within the cell structure. The continuous production of the co-enzymes means that a minor quantity of co-enzymes is utilized very exhaustively (e.g., ATP) [50].

1.6.3 Inhibitor

The enzymatic reactions rate is reduced by different types of enzyme inhibitors [51-53]. There are different types of inhibitors available.

1.6.3.1 Competitive

Competitive inhibitor and reactant substrates bind with enzymes at different times and rates. [54] Generally, competitive inhibitors are approximately identical to reactant substrates e.g., the drug (methotrexate) is approximately identical to dihydrofolate, which is reduced to tetrahydrofolate through dihydrofolate reductase enzymes, meanwhile drug works as a competitive inhibitor of enzymes [55]. The identical structure of the drug helps it to bind with enzymes at the binding site and inhibits its activity by changing its shape [56].

1.6.3.2 Non-competitive

Non-competitive inhibitors attach to enzymes rather than the substrate binding or active site. The reactant substrate even now attaches with enzymes, enzymes regular affinity and K_m value remains constant. So that, inhibitor minimizes the catalytic activity of enzyme and enzymes V_{max} value is decreased [57].

1.6.3.3 Uncompetitive

The uncompetitive inhibitor has no affinity to attach free enzyme, it binds only enzyme-substrate complex. In accompany of the uncompetitive inhibitor, enzyme-substrate complexes are inert [58]. This Kind of inhibition is usually rare [51].

1.6.3.4 Mixed

Mixed inhibitor attaches to an allocated area and substrate binding and inhibitor efficiency are affecting each other. The activity of the enzyme is minimized but not terminated when an inhibitor is attached to the enzyme. Mixed inhibitors do not obey the Michaelis–Menten equation [15].

1.6.3.5 Irreversible

An irreversible inhibitor entirely suppresses the enzyme's activity; in most instances, the irreversible inhibitor binds to the enzyme via a covalent connection. [53]. Penicillin [59] and aspirin [60] are drugs that work on the same principle.

1.6.4 Functions of inhibitors

In most multicellular organisms, an inhibitor acts as a feedback mechanism of the chemical reaction. If an enzyme manufactures a large amount of a single substance in the multicellular organisms, the excessive amount of that substance acts as an inhibitor for this enzyme and decrease or cease the rate of reaction when there the amount of specific substance is sufficient. This kind of feedback is known as negative feedback. Important metabolic channels e.g., the citric acid cycles make utilization of mechanism.

1.7 Mechanism of enzymes working

1.7.1 Substrate binding

For proper working of the enzyme, an enzyme binding to its substrates is compulsory so it can easily catalyze a chemical reaction. Enzymes generally bind with specific substrates that are catalyzed within a specific reaction. This specificity of substance is accomplished by binding sites possessing complementary shapes, charges, and characteristics (hydrophobic or hydrophilic) of the substrate. So, the enzymes can differentiate between identical substrates or molecules due to their specificity in region-selective, chemo-selective, and stereo-specific [40].

A variety of enzymes express the highest accuracy and specificity in repeating and explanation of genome molecules. A lot of enzymes express a "proof-reading" mechanism. For example, DNA polymerase is an enzyme its reaction occurred in 2 phases, in the first phase catalysis of reaction takes place and in the second phase it verifies products, is it correct or not [61]. This two steps reaction results in < 1 error in 100 million reactions. Alike proof-reading mechanism is also the same in aminoacyl tRNA synthetases [62], RNA polymerase [63], and ribosomes [64] [65, 66].

1.7.2 "Lock and key" model

In 1894, Emil Fisher proposed a model to elucidate the distinguish specificity of the enzyme. According to this model, both the substrate and the enzyme have specific

Materials Research Forum LLC
https://doi.org/10.21741/9781644901977-1

compatible geometric appearances that fix completely into each other [67]. This kind of model is known as the " Key and Lock " model. This premature version elaborates enzymes specificity, but cannot elaborate transition states of the enzyme, which are responsible for the catalysis of reaction [68].

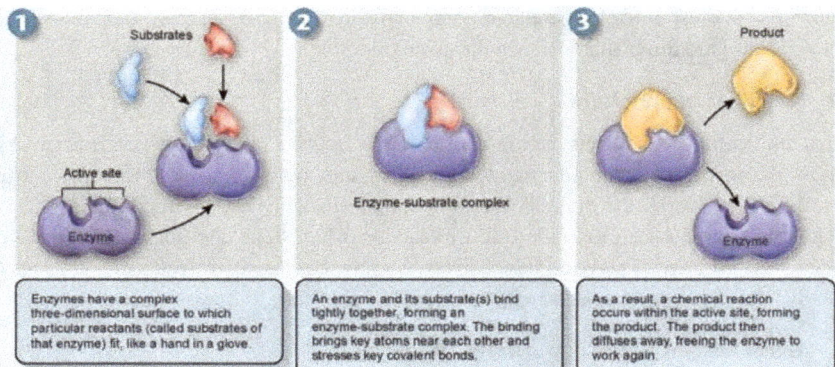

Fig. 2: Mechanism of Enzymes working

1.7.3 "Induced fit" model

In 1958 Daniel Koshland proposed a revision to the "Key and Lock model". As the enzyme's active site is constantly changing due to the interlinking or bonding of substrate with the enzyme, it has a flexible shape [69]. In a conclusion, the substrate cannot join a firm active area; the chains of the amino acids that contain active areas are fouled into an accurate site that authorizes the enzymes to execute the activity. In a few processes, for example, glycosidases, the substrates switch their structure a little bit as it binds to the active site of the enzyme [70]. Changing in the active areas remains continues until substrates are fixed when it is fixed its structure, charge, and properties are verified [71]. Induced Fit Model increases the loyalty of molecular acknowledgment in the existence of competition and proofreading mechanisms [72].

1.7.4 Catalysis

Enzymes can speed up a chemical reaction in many ways, all the mechanisms of enzymes lower the activation energy of chemical reaction (e.g., ΔG^+, Gibbs free energy) [73].

By stabilizing transition states:

Design surrounding having charges distribution compatible to transition states to reduce activation energies [74].

By giving a different channel of chemical reaction:

In the meantime of reaction, the enzyme reacts with substrate, through covalent bonding to supply less energy for transition states [75].

By de-stabilizing the ground states of substrate:

Distorting the bounded substrate to minimize its activation energy necessary for transition state [76].

By aligning substrates into an effective arrangement to minimize [77] the changing in the reaction entropy [78].

1.7.5 Dynamics

Enzymes are not hard, fixed shapes. Enzymes have complicated interior dynamic movements; these movements within enzyme's molecules are due to single amino acid residues, these groups of amino acids result in a unit of secondary design known as protein loop or whole protein domain. These movements bring about a configurational group of a little bit different configurations that replace each other at an equilibrium point [79].

1.7.6 Substrate presentation

This is a procedure in which an enzyme is isolated from specific substrates. An enzyme may be separated from the plasma membrane and isolated from substrates in the cytosol or nucleus. Or inside membranes, enzymes may be separated from lipid, leach out from substrates in chaos area. When enzymes are discharged it combines with substrates.

1.7.7 Allosteric modulation

Allosteric area is known as pockets at enzymes surface, different from active sites, which joins molecules within cellular structure. As result, this molecule is responsible for the alteration of dynamics or conformation of enzymes, which is transformed to the binding area so it influences the rate of reaction of enzymes [80]. Through this mechanism, allosteric interactions may activate or inhibit enzyme activity [81].

1.8   Factor affecting enzymes activity

Enzymes are composed of proteins, they are sensitive to temperature, pH, substrate concentration, changes. The pH optimum for each enzyme is shown in the table below [2, 15].

Table-1: pH optimization for different enzymes

Sr. No.	Enzyme	pH Description	Optimum pH
1	Lipase (stomach)	Acidic	4.0–5.0
2	Arginase	Highly alkaline	10.0
3	Urease	Neutral	7.0
4	Maltase	Acidic	6.1–6.8
5	Amylase (malt)	Acidic	4.6–5.2
6	Ribonuclease	Neutral	7.0–7.5
7	Sucrase	Acidic	6.2
8	Trypsin	Alkaline	7.8–8.7
9	Lipase (castor oil)	Acidic	4.7
10	Cholinesterase	Neutral	7.0
11	Adenosine triphosphate	Alkaline	9.0
12	Catalase	Neutral	7.0
13	Amylase (pancreas)	Acidic-neutral	6.7–7.0
14	Pepsin	Highly acidic	1.5–1.6
15	Cellobiase	Acidic	5.0
16	Lipase (pancreas)	Alkaline	8.0
17	Fumarase	Alkaline	7.8
18	Invertase	Acidic	4.5

1.9   Functions

1.9.1   Biological functions

Enzymes handle a broad range of activities within living organisms. Generally, enzymes are crucial for signal transformation and regulation of cells, frequently through Phosphates and kinases [82]. Enzymes also cause motion, muscle contraction is produced through hydrolyzing ATP with myosin, and also convey shipments all over the cell as a portion of the cytoskeleton [83]. Ion pumps coupled with active transport make up the remaining ATP-ases within the cell membrane. An enzyme can also perform additional remote activities, e.g., luciferase producing light during fireflies [84]. Viruses also accommodate enzymes for influencing cells, e.g., the reverse transcriptase and HIV-integrase, for viral flashes from the cell, like the influenza virus's neuraminidase [85].

The most main activity of the enzyme was found in animals' digestive systems. An enzyme, e.g., proteases and amylases are used to break down larger molecules (proteins or starch) into small molecules so that these molecules are easily absorbed within the intestine. Various types of enzymes digest various types of food molecules. In herbivorous animals that use an herbivorous diet, the microorganism is present in their gut that produces an enzyme "cellulase", which is used to disintegrate the plant's cell wall that is made up of cellulose [86].

1.9.1.1 Metabolism

Several enzymes work simultaneously in a chronic arrangement and establish metabolic pathways. In this type of metabolic pathway, the product of the first enzymes is used as a substrate for the second enzymes and so on. After completion of the first catalytic reaction, a product of the first reaction is then sent to the second enzyme as its substrate. Normally two or more two enzymes can activate the single reaction; this causes more complex regulation: e.g., low rate of chemical reaction activity shown by the first enzyme but second enzyme shows high activity [87].

1.9.1.2 Control activity

For control activity within a cell, there are five main methods. These methods are discussed below.

1.9.1.2.1 Regulation

Enzymes can work as an activator for one substance and at the same time acts as an inhibitor for other molecules. E.g., the metabolic pathway end product generally works as an inhibitor for the first enzyme of this metabolic pathway, so regulating the number of end products formed through this metabolic pathway. Because the concentration of a reaction is controlled by the amount of the same reaction's end product, this type of regulation system is known as a negative feedback mechanism [39].

1.9.1.2.2 Post-translational modification

A few cases of post-translational modifications are myristoylation, phosphorylation, and glycosylation [39]. e.g., to insulin response, multiple enzymes phosphorylation includes glycogen synthase assist to authorize the degradation or formation of glycogen and permits the cell to react with different sugar levels in blood [88]. The breakdown of peptide chains is the second example of post-translational changes. For example, chymotrypsin is a digestive protease that is generated in an inactive state in the pancreas as chymotrypsinogen and then transported to the stomach in the same form and activated in the stomach. This procedure aids in the preservation of pancreatic and other digestive tissue before it enters the intestine. The term "proenzyme" or "zymogen" refers to inactive enzyme precursors.

1.9.1.2.3 Quantity

The formation of enzymes can be enhanced or suppressed through cell responses according to changing of cell environments. This type of gene regulation is called "enzyme induction" e.g., a bacterium becomes resistant to antibiotics like penicillin, the reason is that β-lactamases enzymes are induced, which hydrolyses significant β-lactam rings inside penicillin molecules [89]. Other examples, Enzyme from liver play key role in drugs metabolism is known as cytochrome P450 oxidases, inhabitation or induction of this enzyme may cause the reason of drug interactions [90].

1.9.1.2.4 Subcellular distribution

Some enzymes may be compartmentalized because enzymes have various kinds of metabolic pathways taking place in various cellular portions e.g., fatty acids are formed from one kind of enzyme present in cytosol, Golgi apparatus, and endoplasmic reticulum and consumed by another kind of enzyme as an energy source in the mitochondria by beta-oxidation [91]. Furthermore, transporting enzymes to different parts of the body might alter the oxidative states (e.g., cytoplasm reducing or periplasm oxidizing) or degree of protonation (e.g., acidic lysosomes and neutral cytoplasm), affecting the enzyme's function [92]. Enzyme polymerization into macromolecular cytoplasmic filaments, on the other hand, alters subcellular localization through partitioning into membrane-bound organelles [93, 94].

1.9.1.2.5 Organ specialization

In eukaryotes, multicellular organisms, cells from various tissues and organs have a discrete arrangement of gene expression and consequently, they have a discrete set of the enzyme (called "isozymes") applicable for metabolic reaction e.g., hexokinase is the introductory enzyme in the glycolysis pathway and its specific kind is known as glucokinase exhibited in pancreas and liver [95]. These types of enzymes are hooked in regulating insulin production and sensing blood sugar [96].

1.9.2 Industrial applications

In the chemical industry enzymes are applied and specialized catalysts are required in industrial processes. As a result, proteins engineering is an active research area and it includes its aims to form new enzymes possessing unique properties through in vitro evolution or rational design. [97, 98] These attempts of bioengineering become successful, and several enzymes have been formed "from scratch" to catalyze reaction which is not found in nature. [99]

Materials Research Forum LLC
https://doi.org/10.21741/9781644901977-1

Table-2: Industrial application and utilization of Enzymes

Application	Enzyme Used	Uses	Ref.
Biofuel Industry	Cellulases	Decomposition of cellulose to sugar molecules that make cellulosic ethanol through fermentation.	[100]
	Ligninases	Formation of biofuel through pretreatment of biomass.	[100]
Biological Detergent	Lipases, amylases, proteases	Removal of starch, protein, oil, or fat stains from dishware and laundry.	[101]
	Mannanases	Removal of food stain from the common food additives guar-gum.	[101]
Brewing industry	Proteases, glucanases, amylase,	Break down of proteins and polysaccharides in malt.	[16]
	β-glucanases	Advancement in the worth and beer filtrations properties.	[16]
	pullulanase, Amylo-glucosidase	Production of low calories beer and regulate its fermentation ability.	[16]
	Acetolactate decarboxylase	Enhance fermentation efficiency by suppressing the formation of di-acetyl.	[102]
Culinary application	Papain	Cooking of tenderizing meat.	[103]
Dairy Industry	Rennin	During the production of cheese hydrolyzes protein molecules.	[104]
	Lipases	Produce Camembert cheeses and blue cheeses e.g., Roquefort.	[105]
Food Processing	Amylases	Produce sugars from starch, such as in making high-fructose corn syrup.	[106]
	Proteases	Lower the protein level of flour, as in biscuit-making.	[104]
	Trypsin	Manufacture hypoallergenic baby foods.	[2, 104]
	pectinases, cellulase	Elucidate fruits juice.	[107]
Molecular Biology	Polymerases, DNA ligase , and nucleases,	To produce recombinant DNA via PCR or restriction digestion.	[41]
Paper industry	Xylanases, lignin peroxidases, and hemicellulases	Removal of lignin from kraft pulp.	[108]
Personal Care	Proteases	Removal of protein from contact lens to prevent infection.	[109]
Starch Industry	Amylases	Transform starch into glucose and different syrups.	[110]

2.　Different methods for enzymes immobilization in nanomaterials

The method choice of immobilization is tremendously helpful to restrict deprivation of enzymes activity through no changing of reactive groups at binding sites and the chemical nature of the enzyme. Sufficient knowledge of the nature of active sites is helpful to understand enzymes. Secondly, active sites are secured through binding of protective groups, after complication of reaction these protective are removed without alteration in the activity of enzymes. In a few reactions, this protective activity was achieved through competitive inhibitors or substrates of enzymes. The commonly applicable processes for immobilization of enzymes are covalent coupling, adsorption, cross-linking, and entrapment [111]. Fig. No. 3 represents the pictorial representation of different procedures of immobilization. A lot of review articles explain detailed protocol, methodology, and advantages and disadvantages of each method. A summary of each process is given below.

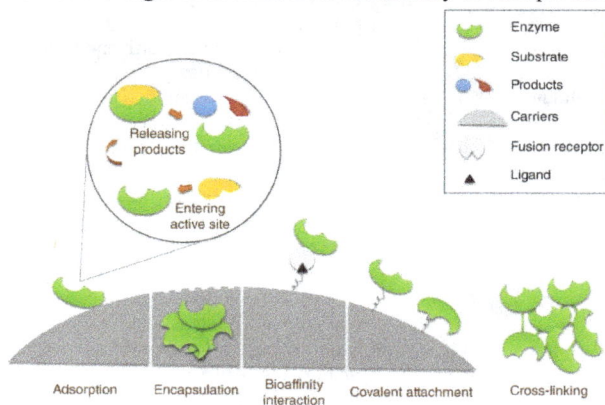

Fig. 3: Graphical view of enzymes immobilized method and reaction of immobilized enzymes [112].

2.1　Adsorption

Adsorption is the traditional, straightforward and easy process of enzymes, this process has a broad range of applications and excessive potential/ability of loading enzymes relative to various immobilization processes. In adsorption, enzymes are generally immobilized through the directly blending of the enzyme with appropriate adsorbent, having a suitable situation of ionic strengths and pH of adsorbent. Unbounded enzymes and loosely attached enzymes are removed through the washing process; the resultant obtained immobilized enzyme is used directly without any further processing. The adsorption method depends upon ionic bonding, Vander Waal forces, hydrophobic interactions, and hydrogen bonding, all these forces are extremely weak, but in huge quantity, impart significant bonding forces [113]. The selection of adsorbents especially relies on decreasing the used enzymes leakage. The adsorption process is carried out by physical techniques cause's a lot of alteration in the protein microenvironment [114]. The disadvantage of this process is enzymes are quickly desorbed by various influences e.g., fluctuation of temperature, alteration of ions, and substrates quantity [115].

2.2   Covalent bonding

Covalent bonding immobilization concerns the evolution of covalent bonds in the support matrix and enzymes. An enzyme contains different types of functional groups, these functional groups are not involved in the chemical activity of enzymes they just take part in the binding of the support matrix with the enzyme. Covalent bonding of support matrix to enzymes, the chain of amino acids containing histidine, aspartic acid, arginine, and reactivity depends upon various functional groups like phenolic hydroxyl, imidazole, indolyl, etc [36]. Modified surfaces of peptide chain applied for enzyme bonding outcomes in higher stability and specific activity having controlled orientation of the protein [116]. CNBr-activated-Sepharose and CNBr-agarose containing glutaraldehyde are spacer arms and carbohydrates are moieties that impart thermal stability to enzymes bounded through a covalent bond [117, 118].

Table 3: Industrial application of Enzymes and their uses and examples represented for covalently bounded enzyme immobilization

Sr. No.	Reaction groups		Type of covalent attachment	Catalyst	Ref.
	Nanoparticles	Enzyme			
1	Carboxyl	Amino	Amino with carboxyl reaction	NHS and carbodiimide	[119, 120]
2	Amino Aldehyde	Carboxyl Amino	Aldehyde with amino reaction	With/without reducing agents	[121, 122]
3	Epoxide functionalized	Amino	Amino with epoxide reaction	Neutral pH	[123]
4	Unsaturated carbonyls (maleimide)	Thiol	Thiol linker	Anatomical pH (6.5–7.5)	[124, 125]

2.3   Entrapment

Entrapment explains as the controlled motion of enzymes in a permeable gel, even treating them as a free molecule in solution. Enzymes traping inside fibers or gels is a suitable procedure, involves I processing of low molecular weight products and substrates. This entrapment technique entirely involves covalent bonding and physical caging. An enzyme entrapped in natural polymers e.g. gelatine, agar, and agarose via thermo reverse polymerization, even in carrageenan and alginate through ionotropic gelation. [126] A large quantity of synthetic polymer e.g., polyacrylamide [127], polyvinyl alcohol hydrogel [128], also has been studied.

2.4   Cross-linking

This process comprises the attachment of biocatalysts with one another through multifunctional or bi-functional ligands or reagents [126]. Through this process, insoluble aggregates having higher molecular weight are produced. As compared to the previous process, the cross-linking process is a comparatively easy method. The cells and enzymes have cross-linked along with inert proteins e.g., albumin, collagen, and gelatine, and may apply to cells or enzymes. The most recently evolved cross-linked enzymes aggregate (CLEAs) is formed through enzymes precipitation from aqueous solution just like protein physical aggregate, through salt addition, non-ionic polymers, and water-miscible organic solvents [129]. CLEAs are easily recycled and show sufficient performance and stability for selective application [130]. The basic process of immobilization of enzymes is categorized into some methods as discussed above, A lot of variation depends upon the union of real procedure, has been evolved [126, 131, 132]. A lot of carriers for various chemical and physical natures of various phenomena' have been proposed for the diversity of bio-separations and bio-immobilizations [126, 133].

2.5 Bio-affinity interactions and other techniques

Many protein-to-small molecule and protein-to-protein interactions, particularly bonding interactions, for example, a histidine-tagged protein with chelated transition metal [134] and biotin-functionalized enzymes, and streptavidin or avidin [135, 136] have been studied and evolved for enzyme immobilization. Immobilization and affinity adsorption on magnetic nanoparticles have also been improved by combining maltose-bonding protein with enzyme [137]. In addition to these immobilization processes, an attainable fusion of these processes also has been evolved to assume specific characteristics for the formation of EnNPs [135, 138].

3.   Enzymes immobilization on different nanomaterial
3.1   Immobilization of carbonaceous nanomaterials

Because of their outstanding qualities, carbonaceous nanomaterials incorporating, graphene, graphene derivatives (e.g., graphene oxide (GO)r and reduced graphene oxide (rGO)), and carbon nanotubes have been published as enzyme immobilization carriers.

This carbon nanomaterial is the leading member of carbonaceous nanomaterial and their application in enzyme immobilization is described in Table 4 and evaluated as under.

3.2 Carbon nanotube

Carbon nanotubes, which come in single-wall and multi-wall varieties, are one-dimensional hollow tubes made up of carbon atoms. CNTs are increasingly being used to immobilize enzymes due to their unique mechanical, thermal, biocompatibility, and electrical capabilities. CNTs have previously been formed via high-pressure carbon monoxide disproportionation, laser ablation, arc discharge, and chemical vapor deposition [139]. Feng's group administrated research on enzymes immobilization along MWCNTs, [140, 141] which provide some knowledge of activation's system of MWCNTs immobilized enzyme [142].

3.2.1 Graphene

Graphene is a two-dimensional honeycomb-like network of carbon atoms, and one-atom-thick, that has fascinated growing attention in a variety of swiftly evolving disciplines because of its elevated surface area, great mechanical strength, and noteworthy thermal, optical, and electrical properties [143]. Graphene assumes to be a perfect carrier for the immobilization of enzymes due to its extraordinary features. The larger surface area assists to raise enzyme packing and the glowing mechanical strength enlarges the enzyme recyclable.

3.2.2 Graphene oxide and reduced graphene oxide

Owing to the extraordinarily larger surface area, outstanding mechanical, and thermal qualities, graphene oxide is known as the perfect immobilized enzymes carrier. Graphene oxide possesses a lot of oxygen and diverse functional groups including epoxy, carboxyl, and hydroxyl, which makes it ideal for enzyme immobilization. The Hummers method has been used for the fabrication of graphene from graphite oxidation [144]. Graphene oxide immobilized enzyme has exhibited various catalytic activities when Graphene oxide nano support is developed through various procedures. Reduced graphene oxide has lost its functional group on its surface, which is attained during graphene oxide reduction. Covalently connected enzymes, electrostatically adsorbed or cross-linked enzymes, and hydrophobically adsorbed enzymes on reduced graphene oxide are all possible. Physical adsorption, covalent attachment, and crosslinking are used to construct the graphene oxide immobilized lipase by Hermanová et al. [145].

3.3 Immobilization on metal/metal oxides nanomaterials

A metallic nanomaterial is a division of nanomaterial consisting of metal elements. The application of metals, metal oxides, and metal hydroxides nanomaterial in the immobilization of enzymes is discussed.

Table 4: The applications of Au nanomaterial immobilized enzymes

Sr. No.	Enzyme	Nanomaterial	Immobilization method	Application	Ref.
1	HRP	AuNPs	Chemical adsorption	Biosensor	[148]
2	Periodate-oxidized GOD	AuNPs-NH$_2$	Schiff base covalent coupling	Biosensor	[149]
3	Periodate-oxidized GOD	AuNPs-NH$_2$	Schiff base covalent coupling	Biosensor	[152]
4	GOD	AuNPs-COOH	EDC/NHS covalent coupling	Thermo-stability improvement	[119]
5	GOD	AuNPs-COOH	EDC/NHS covalent coupling	Estimation of glucose through eye	[153]
6	Cholesterol oxidase	AuNPs-COOH	EDC/NHS covalent coupling	Biosensor for cholesterol estimation in blood stream	[154]
7	GOD	AuNPs	GA crosslinking	Biosensor	[155]
8	α-Amylase	AuNPs	Ionic exchange/ Hydrophobic interaction	Starch processing	[156]
9	Cholesterol oxidase	AuNPs-COOH	EDC/NHS covalent coupling	Biosensor for cholesterol estimation in blood stream	[157]
10	HRP	AuNPs-NH$_2$	GA crosslinking	Biosensor	[150]

3.3.1 Metal nanomaterial

Because of its higher surface area, virtuous mechanical and thermal stability, simple functionalization, and biocompatibility, the gold nanomaterial is a popular choice for immobilized enzymes among metal nanomaterials. Gold nanomaterial has been formed through citrate two-phase reduction process, reduction process; seed-mediated evolution process, etc., all these processes have been investigated by Dreaden and his coworkers [146]. The enzyme is immobilized on nude or gold functionalized nanomaterial through physical adsorption [147], chemical adsorption [148], covalent attachment [149], or cross-linking [150]. Apart from this, the gold nanomaterial is always looked at with other nanocarriers for the immobilization of enzymes [151]. The gold nanomaterials immobilized enzyme is always used in biosensors and their achievements are described below in Table 4.

3.3.2 Metal hydroxide

Layered double hydroxide (LDHs) is a type of metal hydroxide nanomaterial with a layered architecture that has gained a lot of attention for its usage in the drug delivery process [158], catalysis [159], treatment of wastewater [160], and many more. In the past few years, because of their greater surface area and excellent biocompatibility, LDHs have been created as a suitable substrate for immobilized enzymes [161]. The enzyme may be settled through ion exchange, electrostatic attraction, and hydrogen bonding or van der Waals interaction [162].

3.3.3 Metal oxide nanomaterials

In metal oxide nanomaterial, TiO_2 [163], Fe_3O_4 [164, 165], and ZnO [166] nanomaterials are generally utilized for enzymes immobilization. The procedure for the formation and functionalization of Fe_3O_4 nanoparticles and their utilization in enzymes immobilization has been extensively reported. [167, 168] ZnO and TiO_2 nanomaterials are also used as a carrier of enzyme immobilization.

3.4 Immobilization of conductive polymers

Due to its appealing qualities for the evolution of biosensors, conducting polymer has gotten a lot of interest [169]. High electrical conductivity, low ionization potential, high electronic affinity, and outstanding optical qualities characterize conducting polymers. The most applicable systems depend on polyaniline, polyacetylene, poly-pyrrole, poly, polythiophene, and polyphenylene, vinyl, or -ene. Outstanding conducting polymers dependent enzymes biosensor has been produced for estimation of biological molecules e.g., acetylcholine, [170] glutamic acid [171, 172], and soluble gasses e.g., NO [173], all of these linked with specific diseases.

3.5 Enzyme immobilization on other materials

Apart from these nanocarriers mentioned above, additional nanomaterials e.g. chitosan [174], silica [175], and cellulose [176] -based nanomaterials, mesoporous materials [177], clay minerals [178], and metal-organic framework [179], are used as immobilization of enzymes. An et al. provide a summary on applications of clay minerals as a carrier of immobilized enzymes. The enzymes immobilized on structured mesoporous material and their use was explored by Zhou and Hartmann. [180]. Lian et al. looked into using a metal-organic framework as an enzyme carrier for immobilization [181].

4. Application of immobilized enzymes on nanomaterials

4.1 Electrochemical sensing applications of enzyme immobilized on nanomaterials

An electrochemicalcal biosensor is a broadly applicable biosensor whose mechanism of working depends upon electrochemical features of analytes and transducers [182]. Alteration in the features of physicochemical of electro-active molecules e.g., resistance, voltage, current or superficial charge, formed through redox reaction caring on transducer surface of the electrode, is output signal. The application of electrochemical biosensors produces advantages e.g., rapidity, simplicity, high sensitivity, and low cost. Potentiometry, amperometry, impedimetric, and conductometry are typical forms of transducers applied in electrochemical biosensors.

4.1.1 Amperometric biosensors

The estimated value in enzyme-based amperometric biosensors is the current created by the reduction or oxidation of electro-active substances at their electrodes (i.e., platinum, carbon, gold, etc.). When a prescribed potential is altered in the middle of two electrodes, the amount of produced current at the working electrode's surface is proportionate to the analyte quantity of available in trail solutions. The advantages of amperometric biosensors include their resilience, ease of downsizing, and ability to manage complex matrices with small sample volumes [183, 184]. Based on the electron transfer technique, which is used to estimate biological reactions, these biosensors have evolved through three generations [183, 184]. Various types of enzyme-based amperometric biosensors are now available for estimating alcohol, lactate, glucose, and other compounds by exploiting oxidases (e.g., alcohol oxidase, lactate oxidase, glucose oxidase, etc.) to oxidize their substrate and produce hydrogen peroxide (H_2O_2), which is measured through the electrode.

4.1.2 Potentiometric biosensors

Being a variety of enzymes are related to the absorption or release of hydrogen ions during the enzymatic reaction, and its outcome is variation in ions amount, ionic selective electrodes may be applied to detect this techniques variation in voltages between reference electrodes and working electrodes is known as a signal in potentiometric biosensors, and it is computed under equilibrium conditions. The computed signal is used for quantification of analyte amount [185]. Classification of Potentiometric biosensors is light-addressable potentiometric sensors, enzyme field-effect transistors, and Ion-selective electrodes.

4.1.2.1 Ion selective electrode

As a potentiometric sensor, an ion-selective electrode is utilized; ionic selective electrodes convert the activity of certain ions in sample solutions into potential (voltage), which is monitored using a pH/mV meter. These electrodes generally consist of two constituents:

- An ions selective membrane that gives permeability for selective ion in sample solution of analyte that contains several interfering ions,
- An integrated or separate reference electrode.

4.1.2.2 Enzyme field-effect transistors

On ion-sensitive field-effect transistors constructed by dividing the metal gate of conventional metal oxides, an enzyme field-effect transistor sensor is produced, removing the semiconductor field-effect transistor from the device and reinstalling the gate in the form of reference electrodes dipped in water solutions separated from the gate oxide by enzymatic membranes [186].

4.1.2.3 Light addressable potentiometric sensors

Rather than using alternating current voltages to activate the semiconductor in light addressable potentiometric sensors, a modulated light-produced light-emitting diode is used. Electron-hole pairs form on the semiconductor's surface as a result of lightning. As a result of the fixed bias voltages, a photocurrent is established and estimated. The light addressable potentiometric sensors are semiconductor-dependent chemical sensors that produced the electrolyte insulator semiconductor form [187].

4.1.3 Conductometry

Enzymes reaction is generally associated with variation in the concentration of ions, which changes the conductivity of electrolytic solutions. The conductivity of the solution is evaluated by conductometric biosensors via implementing potential voltages across two parallel dipped electrodes. As a result, the mobility of negative charge ions approaching the anode and positive charge ions approaching the cathode increases ionic motility. The electrolytic solutions conductivity is purely due to mobility and ion concentration. The enzyme is immobilized at the electrode's surface through sol-gel entrapment technique [188], covalent bonding among collagen membranes [189], electrochemical polymerization technique [190], or cross-linking technique with bovine serum albumin applying glutaraldehyde [191]. The many types of immobilized enzymes and substrates are detailed in Table 5, and all of these immobilized processes are for conductometric enzymes biosensors.

Table-5: Summary of substrates, immobilization enzyme, and immobilized procedures of conductometric enzymes biosensor

Sr. No.	Immobilized Enzymes	Substrates	Immobilized techniques	Ref.
1	Uricase	Uric acid	-	[192]
2	Creatinine deiminase	Creatinine	Adsorption	[193]
3	AChE, BChE	Organophosphates	Cross-linking (GTA)	[194, 195]
4	GOx	Glucose	Covalent bonding (EDC-NHS)	[196]
5	Alcohol oxidase	Formaldehyde	Cross-linking (GTA)	[197]
6	β-galactosidase, GOx	Lactose	Cross-linking (GTA)	[198]
7	D-amino acid oxidase	D-amino acids	Covalent bonding (Hydrazine)	[189]
8	Tyrosinase	Triazine herbicides	Cross-linking (GTA)	[199]
9	Urease	Urea	Cross-linking (GTA)	[200]
10	AChE	Acetylcholine	Cross-linking (GTA)	[201, 202]
11	BChE	Butyrylcholine	Cross-linking (GTA)	[201, 202]
12	Urease	Heavy metal ions	Cross-linking (GTA)	[203]
13	AChE	Carbamate pesticides	Cross-linking (GTA)	[204]
14	Peroxidase	Hydrogen peroxide	Cross-linking (GTA)	[205]
15	L-asparaginase	L-asparagine	Cross-linking (GTA)	[206]

4.1.4 Impedimetric enzyme biosensors

In electro-chemical impedimetric dependent biosensors, the electrode impedance is quantitative value. Electrochemical impedance spectroscopy is applied for evaluating variation in interfacial characteristics, having the property of bio recognitions at enzyme-modified surfaces. The resultant impedance spectra are applied for the estimation of the quantitative criteria of electrochemical techniques. In enzymes based biosensors, the impedance estimation process is not generally the same as in amperometric and potentiometric techniques, because of the consumption of time for recording complete impedance spectra to win a wide range of frequencies. Along with this, in the EIS process, many requirements e.g., causality, stability, and linearity are necessary to get valid impedance spectra. Hence, the EIS method is generally applicable as a characterizable technique for a lot of enzymes based impedimetric biosensors.

Materials Research Forum LLC
https://doi.org/10.21741/9781644901977-1

Table 6: An overview of immobilized enzyme, analyte, immobilized technique, and Limit of Detections of impedimetric enzymes biosensor

Sr. No.	Immobilized Enzymes	Analytes	Immobilization Methods	LOD	Ref.
1	Alcohol oxidase	Alcohol	Electrochemical polymerization (aniline)	-	[207]
2	GOx	Glucose	Covalent bonding (EDC-NHS)	15.6 µM	[208]
3	GOx	Glucose	Adsorption	1 mM	[209]
4	Butyrylcholinest erase	Trichlorfon	Adsorption	0.1 ppm	[210]
5	Glycerol catalase	Cyanide	Photopolymerization (PVA-SbQ)	4 µM	[211]
6	Urease	Urea	Covalent bonding (Eudragit S-100, carbodiimide)	0.02 M	[212]
7	Lipase	Diazinon	Cross linking (GTA)	10 nM	[213]
8	CAT	H_2O_2	Adsorption	0.025 nM	[214]
9	GLOD	Glutamate	Cross linking (GTA)	20 µM	[215]

4.2   Fuel cell applications of enzyme immobilized on nanomaterials

For the generation of biofuel, a variety of nanomaterials have been used. At the activator size, covalently attached immobilized lipase has been implemented on magnetic Fe_3O_4 nanoparticles for practice in the production of biodiesel [216]. Lipases from Candida rugosa, Pseudomonas cepacia, and Porcine pancreas were immobilized on functionalized amino-Fe_3O_4 nanoparticles [217].

The highest production of biodiesel up to 90% was reported from covalent immobilization Thermo-myces lanuginosus lipase with the functionalized amino-magnetic nanoparticle. [165, 218]   Pseudomonas cepacia lipase has been immobilized through adsorption technique on nano-pores gold particles support [219]. Highly expensive gold practicals [220] have been widely applicable for operation because of chemical stability and biocompatibility. [221] PAN nanofibres were activated through amidination reactions for covalently immobilized PCL enzymes [222].

In all the different types of functions of nanomaterial bound enzymes, bio-catalytic efficiency in the production of biofuel has been evolved; a crucial step towards, later application in the formation of biofuel.

Materials Research Forum LLC
https://doi.org/10.21741/9781644901977-1

Table 7: Biosensor applications of few immobilized enzymes nanoparticle

Sr. No.	Nanoparticles	Type of biosensors	Typical application	Properties used	Ref.
1	Glucose oxidase (GOx)-immobilized silver nanoparticles	Optical	Potential test kits for estimation of glucose level	Change in localized surface plasmon resonance with glucose – Simplifying and inducing optical sensitivity	[244]
2	Cholesterol oxidase (ChOx) aggregate	Electrochemical	Amperometric biosensors for estimation of cholesterol level in human blood serum	Change in electrical properties with cholesterol – Increasing stable, active, and shelf life for immobilized enzymes	[245]
3	GOx–gold nanoparticles	Electrochemical and optical	Optical biosensor and electrochemical biosensor in the estimation of glucose Level	Transduction in electrochemical impedance spectroscopy and surface plasmon resonance – Enhancing sensitive response and good reproducibility	[234]
4	β-Galactosidasei mmobilized gold nanoparticles	Colorimetric	Visual test strip format in the estimation of microbial contamination (e.g., model analyte *Escherichia coli*)	Colorimetric development through released enzymes Fast and instant Measurement	[246]

4.3   Bio-sensor applications of enzyme immobilized on nanomaterials

Biosensors are applied for the estimation of an analyte along biological detection components attached with physic-chemical detectors that transform detected responses to quantifiable signals that are proportional to the amount of analyte [223-231]. Enzymes dependent sensors are divided into the following types based on signal-transducing format e.g., calorimetric/thermal, optical, electrochemical, and piezoelectric biosensors. The nanoparticle is also applied as a carrier for enzymes immobilization on biosensors electrodes are summarized in Table no. 7. Silver, Gold, and metals-like nanoparticle

Materials Research Forum LLC
https://doi.org/10.21741/9781644901977-1

outstandingly enhance the enzyme's layered immobilization electrical conductivity at electrodes, which increases the detection sensitivity [232, 233]. AuNPs can also be used as a catalyst to speed up electrochemical reactions, such as the reduction and oxidation of H_2O_2 in glucose biosensors. [234-236]. Nowadays, a large quantity of nanoparticles electrochemical biosensors dependent upon the fusion of magnetic nanoparticles and different other materials e.g., chitosan, electro-conductive polymer, and carbon nanotube has been applied as an electro-catalytic-magneto-switchable biosensor [237-242]. Generally, applying enzymes dependent immobilized nanoparticle in biosensor enhance their analytical activity detected through the dynamic parameter of the biosensors e.g., depression in response time and elevation insensitivity [231, 233, 234, 239, 243]. Many biosensors devices are linked with many enzymes e.g., urease, HRP, ChOx, GOx, acylase, penicillin, and nanoparticle has been commercially used for biomedical, clinical, industrial, environmental, and pharmaceutical analysis [231, 233, 243].

4.4 Enzyme nanoparticles for biomedical application

Enzymes have a large tendency in the therapy of oncological [247], cardiovascular [248], intestinal [249], viral and hereditary [250], and other disorders [251-263]. The routine therapeutic application of enzymes, on the other hand, is not commonly used due to their short lifespan in the human body or for storage, which increases the danger of toxic and systemic immunological adverse effects. These shortcomings can be mitigated by delivering therapeutic enzymes to infected locations in a targeted manner [247]. A large quantity of therapeutic enzymes nanoparticles has to express their role in vitro and animal researches [247].

Table 8: Selected immobilized enzymes and their medical use.

Sr. No.	EC Number	Enzyme	Application
1	4.3.1.5	Phenylalanine ammonia-lyase	Phenylketonurea
2	1.1.1.49	Glucose-6-phosphate dehydrogenase	Glucose-6-phosphate dehydrogenase deficiency
3	1.1.3.22	Xanthine oxidase	Lesch–Nyhan disease
4	4.2.1.1	Carbonic dehydratase	Artificial lungs
5	1.7.3.3	Urate oxidase	Hyperuricemia
6	3.5.1.1	Asparginase	Leukemia
7	1.1.3.4	Glucose oxidase	Artificial pancreas
8	1.11.1.6	Catalase	Acatalasemia
9	4.2.2.7	Heparinase	Extracorporeal therapy procedures
10	3.5.3.1	Arginase	Cancer
11	3.5.1.5	Urease	Artificial kidney
12	3.2.1.3	Glucoamylase	Glycogen storage disease

4.4.1 Thrombolytic therapy

Streptokinase, tissue-type plasminogen activators, and urokinase-type plasminogen activators have all been used to inhibit blood clotting in critical cerebral micro thrombosis or myocardial infarction [248, 249]. Nanoparticle e.g., polymeric nanoparticle, liposome, magnetic nanoparticle, or red blood cell has been selected as carrying this thrombolytic enzyme to the site of blood clotting to abolish the possibility of hemorrhagic reaction produced by nonspecific activation or non-targeting of this thrombolytic enzyme [250-253]. To achieve ultrasound-assisted thrombolysis, tissue-type plasminogen activators are additionally adsorbed into echogenic liposomes [254].

4.4.2 Oxidative stress and tnflammation therapy

Reactive oxygen species consist of a wide range of free radicals of oxygen that is sensitive and can induce oxidative stress on cell structure [255]. Reactive oxygen species from both non-enzymatically and enzymatically, causing cell death in macrophages, epithelial cells, neutrophils, monocytes, eosinophils, and lymphocytes via electron transfer processes in response to a variety of stimuli [255-257]. Intracellular reactive oxygen species creation occurs in innate immune cells at sites of inflammation connected to chronic inflammatory sickness or in areas where oxygen and other exogenous mediums have a strong influence, such as cigarette smoking in the lungs [256]. Cells are primarily guarded against oxidative stress and inflammation by catalytic reactions accomplished via catalase [135, 258] and superoxide dismutase [259]. A new approach for SOD intracellular delivery [260-263] based on cell-penetrating peptides fusion of SOD implanted in mesoporous silica nanoparticles has also been developed recently [264].

4.4.3 Antibacterial treatment

A lot of recent researches have shown the utilization of enzymes coupled nanoparticle for antibacterial application for minimizing the antibiotic resistances rate for human pathogens e.g., E. faecium, M. tuberculosis, P. aeruginosa, and S. aureus [265-267].

Lysozymes were attached to polystyrene nanoparticles and utilized to boost the transport of infested macrophages to the liver via the reticuloendothelial system.

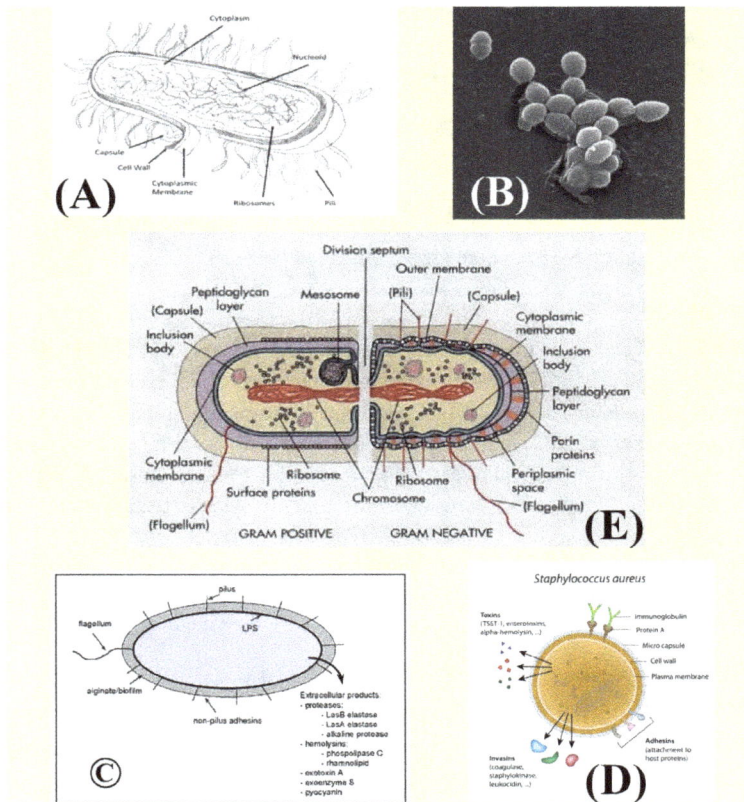

Fig. 4: Structures of pathogenesis (A) M. tuberculosis (B) E. faecium (C) P. aeruginosa (D) S. aureus (E) Difference of Gram-positive and negative bacteria

4.5 Water contaminants treatment applications of enzyme immobilized on nanomaterials

4.5.1 Removal of emerging content

Materialized pollutants are universal in soft and drinking water and generally laborious to abolish in ordinary water treatment techniques e.g., filtration, sedimentation, and coagulation. To enhance the eliminating efficiency of these pollutants, various latest processes have been evolved and applied, including electrochemical oxidation technique, photocatalysis technique, advanced chemical oxidation technique, and enzymatic catalysis technique. Among all of these techniques, the enzymatic technique exhibits comparatively high versatility and efficiency, needs less energy and chemical consumption, and also has a lessened capacity to produce toxic byproducts, which accumulatively makes it a satisfactory process particularly for the POU system [268-270]. Various enzymes e.g., laccase, peroxidases, atrazine chlorohydrolase, and organophosphate hydrolase have been recognized and devised to treat numerous pollutants, e.g., pharmaceutical, endocrine disruptors, personal care products, and pesticides. Pathway, kinetic, and mechanism of enzymatic treatment have been considerably studied [268, 271, 272].

4.5.2 Disinfection

Present drinking water disinfection techniques e.g., ozonation and chlorination produce poisonous byproductsAntimicrobial enzymes, on the other hand, prevent the formation of any harmful byproducts by disrupting bacterial biofilms and microorganisms' cellular component or adhesion [293]. Lysostaphin and lysozyme is powerful enzyme across Gram-positive bacteria reason is they break down cross-linking bonds present in peptidoglycan [275, 279]. Another method can be used for the enzyme, which creates oxidative stressors such as hydrogen peroxide to prevent all types of bacterial development [293, 294]. Immobilized antibacterial enzyme enhances their stability and secure them from being infected through microbes [275].

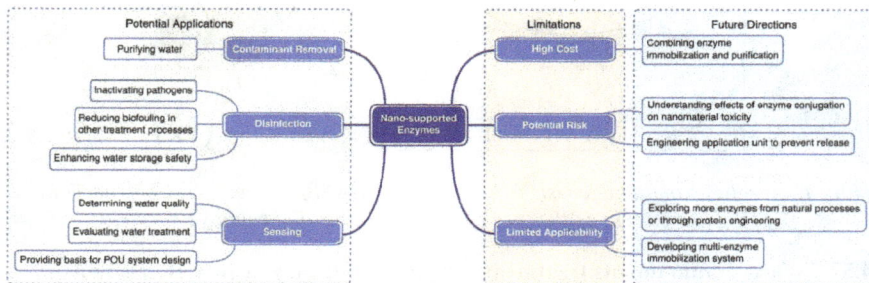

Fig. 5: A summary of the suitability of nano-supported enzyme in POU water system: limitation, potential application, and future direction [295].

Table-8: An overview of Nano-supported metal particles and their Potential Applications and Advantages in the POU System

Nanomaterial	Advantages	Potential Applications	Refs
Silica Nano-supports			
Chemically synthesized silica	Easy synthesis	Contaminant Removal Disinfection	[273-275]
Biosynthesized silica	Milder synthesis conditions less enzyme activity loss	Contaminant Removal Disinfection	
Carbon nano-supports			
CNTs, graphene, and mesoporous carbon	Extraordinary electric conductivity	Enzymatic amperometric Sensor	[276-279]
Nano-gel	Minimal enzyme activity loss	Disinfection, Contaminant removal	[280]
Metallic nano-supports			
MOF and protein−inorganic Nano-flower	Increased enzyme activity	Contaminant removal	[281-283]
Magnetic nanoparticle	Easy recovery	Contaminant removal	[284]
Noble metal nanoparticle	Unique aggregation-state dependent optical property enhanced quantum efficiency of surface-bound fluorophore	Enzymatic colorimetric sensors	[246, 285, 286]
Bio-derived nano-supports			
Protein cage	Environmentally friendly, biocompatible, a programmable structure that allows enhanced enzyme selectivity and Activity	Contaminant removal, disinfection	[270, 287-290]
DNA origami	The biocompatible, enhanced overall activity of enzyme couples	Contaminant removal	[291, 292]

Materials Research Forum LLC
https://doi.org/10.21741/9781644901977-1

4.6 Water contaminants monitoring applications of enzyme immobilized on nanomaterials

Conventional physical technique to monitor pollutants is lengthy, expensive and a lack of instant estimation increases health risk. As a result, improving the sensor to provide fast on-the-spot pollutant assessment can effectively remove the risk of unintentional or incidental disclosure to communities, as well as develop rules for constructing a POU system to meet specific treatment requirements. The reason is this, the enzyme has the superiority of high specificity, sensitivity, and instant response to the substrate, and perform a promising technique for the evolution of biosensor. Nano-support provides the framework required to form enzymes dependent sensors.

4.6.1 Bacterial approach

For bacterial detections, the nano supports interact with the surface of the cell although immobilized enzymes produce a signal. β-Galactosidase catalyze chlorophenol red beta-D-galactopyranoside hydrolysis, chromogenic reactions result in the alteration color. The anionic -galactosidase is absorbed by gold nanoparticles that have been modified with the cationic ligand. Immobilized -galactosidase is used to catalyze a reaction that yields colorimetric yield in conjunction with chlorophenol red -D-galactopyranoside during the adhesion of anionic surfaces of bacteria to the nanoparticle [246].

4.6.2 Colorimetric approach

Colorimetric techniques are applied for the instant detection of pollutants. Contrast with electro enzymatic sensor, minor sensitive but more appropriate for application by non-trained people. The formation of colored compounds from enzymes mediated pollutant transformation can be immediately applied for quantifying concentration [281, 296]. Noble metal nanomaterials' unique optical and surface features are also crucial in the development of colorimetric enzyme sensors. The quantum performance of surface constrained fluorophores can be improved by using nano-metal. A fluorophore is a competitive enzyme inhibitor that can be attached to immobilized enzymes and retained on the nano-metals surface, resulting in increased fluorescence.

4.6.3 Electro-enzymatic approach

The electro-enzymatic biosensor is made up of enzymes that are coated on the electrode and can be used to estimate contaminants. When an electric potential is produced, molecules formed or consumed by enzymatic activity produce current, as well as activity due to pollution levels. Although there are restrictions to the activity of these types of sensors, such as the destruction of enzyme activities and stability, and the electron transfer barrier formed by enzyme covering on the electrode surface, there are advantages to using them [297].

4.7 Other applications of immobilized enzymes on nanomaterials

Other uses for immobilized enzymes can be noticed in the fabric industry, whereas the major advantage is their cost-effectiveness. The size of the manufacturing facility required for constant processing is two times lesser than that necessary for batch processing with the free enzyme. The whole cost is, too much smaller. Immobilized enzyme offers appreciably enhanced activity depending upon enzymes load and also usually gives processing advantage [298].

Besides that, the oxidation reaction is the principal reaction in numerous industries; mostly applies the traditional oxidation process, which has the following disadvantages: undesirable or non-specific side reaction and utilization of environmentally hazardous chemicals. These drawbacks have urged the search of the latest oxidation processes depending on the biological system e.g., enzymatic oxidations. This system expresses the following improvement over chemical oxidations: an enzyme is a biodegradable and specific catalyst and enzymes reaction is conducted under normal settings.

Enzymatic oxidation has been proposed in a range of industries, comprising foodstuff, fabrics, and pulp & paper [299]. The oxidation of ortho and para amino phenols, di-phenols, polyamines, polyphenols, aryl di-amines, lignins, and a few inorganic ions, as well as the decrease of molecular di-oxygen to water, is catalyzed by these enzymes [300].

A lot of different kinds of procedures and supports are continuously applied for immobilized numerous industrial enzymes, and a lot of functions of laccase have been suggested [301]. Laccase has recently become essential in the textile sector for decolorization of dye processes, as well as in the pulp and paper business for dignifying woody fiber, particularly during the bleaching process [302]. Laccase is utilized in conjunction with a chemical mediator in all of these applications [303].

Conclusion

Latest advancements in the advancement of nanostructured supported material and immobilized techniques have authorized accurate immobilization of biocatalyst, so that, in recent they have got more importance in flourishing the nano-biocatalyst-mediated industrial bio-processing. Especially, functionalized nanomaterial extremely affects the inherent mechanical properties and provides the highest biocompatibility and specific nano-environment surrounding the enzymes for improving enzymes stability, catalytic performance, and reaction's activities. Moreover, immobilized enzymes on nanocarriers composed can considerably enhance the robustness and durability of the enzyme for its frequent applications, which reduces the overall price of the bio-catalytic process. Consequently, increasing fundamental understanding and knowledge of the physicochemical characteristics of biocatalysts and nanocarriers, additionally, and interfacial interconnection associated with the nanocarriers-enzymes systems. Rather than all these improvements, various problems are still present that requires controlling for real-time application at an industrial scale. Firstly, the price is linked with enzymes preparation or bio-synthesis, nanostructured carriers, and immobilized enzymes are the main problems for industrial-scale application. On the other hand, a large quantity of nanomaterials is still produced at the laboratories scale. Secondly, a lot of nano-biocatalysts evolved only to express proof of concept realization, and a small number of nanocarriers immobilized biocatalyst has been profitably managed for a maximum of 10 to 15 consecutive repeated batches. So that durability, catalytic activity, leaching of the enzyme, and mechanical steadiness are evaluated for their continual operation.

References

[1] D. Schomburg, M. Salzmann, Enzyme handbook, Enzyme Handbook, Springer1991, pp. 1-1175. https://doi.org/10.1007/978-3-642-76729-6_1

[2] E. Buchner, " Biocatalyst" redirects here. For the use of natural catalysts in organic chemistry, see Biocatalysis.

[3] J.M. Murphy, H. Farhan, P.A. Eyers, Bio-Zombie: the rise of pseudoenzymes in biology, Biochemical Society Transactions 45 (2017) 537-544. https://doi.org/10.1042/BST20160400

[4] J.M. Murphy, Q. Zhang, S.N. Young, M.L. Reese, F.P. Bailey, P.A. Eyers, D. Ungureanu, H. Hammaren, O. Silvennoinen, L.N. Varghese, A robust methodology to subclassify pseudokinases based on their nucleotide-binding properties, Biochemical Journal 457 (2014) 323-334. https://doi.org/10.1042/BJ20131174

[5] I. Schomburg, A. Chang, S. Placzek, C. Söhngen, M. Rother, M. Lang, C. Munaretto, S. Ulas, M. Stelzer, A. Grote, BRENDA in 2013: integrated reactions, kinetic data, enzyme function data, improved disease classification: new options and contents in BRENDA, Nucleic acids research 41 (2012) D764-D772. https://doi.org/10.1093/nar/gks1049

[6] A. Radzicka, R. Wolfenden, Rates of uncatalyzed peptide bond hydrolysis in neutral solution and the transition state affinities of proteases, Journal of the American Chemical Society 118 (1996) 6105-6109. https://doi.org/10.1021/ja954077c

[7] B.P. Callahan, B.G. Miller, OMP decarboxylase-An enigma persists, Bioorganic chemistry 35 (2007) 465-469. https://doi.org/10.1016/j.bioorg.2007.07.004

[8] S. Paroha, R.D. Dubey, S. Mallick, Recent Targets in Drug Discovery: A Review, Research Journal of Pharmacology and Pharmacodynamics 3 (2011) 5-9.

[9] B.E. Güler, Farklı aminoasit içeren nanofibrillerde bacillus subtılıs E6-5 suşundan ham proteaz ve ticari proteazın immobilizasyon çalışmaları, Uludağ Üniversitesi, 2017.

[10] M.B. Duza, S. Mastan, Microbial enzymes and their applications-a review, American Journal of Pharm Research 3 (2013) 6208-6219.

[11] A. Payen, J.-F. Persoz, Mémoire sur la diastase, les principaux produits de ses réactions, et leurs applications aux arts industriels, Ann. chim. phys 53 (1833) 73-92.

[12] K.L. Manchester, Louis Pasteur (1822-1895)-chance and the prepared mind, Trends in biotechnology 13 (1995) 511-515. https://doi.org/10.1016/S0167-7799(00)89014-9

[13] A.F. Agrò, G. Mei, Allostery: The Rebound of Proteins, Allostery, Springer2021, pp. 1-6. https://doi.org/10.1007/978-1-0716-1154-8_1

[14] E. Buchner, Academic Dictionaries and Encyclopedias.

[15] E. Buchner, Etymology and history.

[16] E. Buchner, From Infogalactic: the planetary knowledge core Jump to: navigation, search" Biocatalyst" redirects here. For the use of natural catalysts in organic chemistry, see Biocatalysis.

[17] M.A. Fraatz, M. Rühl, H. Zorn, Food and feed enzymes, Biotechnology of Food and Feed Additives (2013) 229-256. https://doi.org/10.1007/10_2013_235

[18] O. May, Industrial Enzyme Applications-Overview and Historic Perspective, Industrial enzyme applications (2019) 1-24. https://doi.org/10.1002/9783527813780.ch1_1

[19] M. Piccolino, Biological machines: from mills to molecules, Nature Reviews Molecular Cell Biology 1 (2000) 149-152. https://doi.org/10.1038/35040097

[20] V. Politi, Specialisation and the incommensurability among scientific specialties, Journal for General Philosophy of Science 50 (2019) 129-144. https://doi.org/10.1007/s10838-018-9432-1

[21] L. Shelfer, The alchemy of jargon: Etymologies of urologic neologisms. Number 2: Basic biochemical nomenclature, Wiley Online Library, 2009. https://doi.org/10.1002/pros.20876

[22] J.L. Heilbron, The Oxford companion to the history of modern science, Oxford University Press2003.

[23] M.P.C. Bernal, The translation of scientific literature from German into Spanish at the turn of the 20th century, Translation & Interpreting 11 (2019) 87-105. https://doi.org/10.12807/ti.111202.2019.a08

[24] Marquis, R.E., W.P. Brown, and W.O. Fenn, Pressure sensitivity of streptococcal growth in relation to catabolism. Journal of Bacteriology, 1971. 105(2): p. 504-511.

[25] R.B. Merrifield, Robert Bruce Merrifield, Nobel Prize in Chemistry 393.

[26] R.B. Merrifield, HOME• META SEARCH• TRANSLATE.

[27] Moerner, W.E., High-resolution optical spectroscopy of single molecules in solids. Accounts of chemical research, 1996. 29(12): p. 563-571

[28] Kador, L., Optical detection and Spectroscopy of single molecules in a solid. Phys. Rev. Lett, 1989. 62(21): p. 2535-2538.

[29] Seaborg, G.T., History of Met Lab Section CI. 1980.

[30] E. Buchner, Nobel Lecture: Cell-Free Fermentation, Nobelprize. org (1907).

[31] C.-L. Liu, Relocating Pastorian Medicine: Accommodation and Acclimatization of Medical Practices at the Pasteur Institutes in China, 1899-1951, UCLA, 2016.

[32] N.J. Mulder, Protein family databases, eLS (2007). https://doi.org/10.1002/9780470015902.a0003058.pub2

[33] C.B. Anfinsen, Studies on the principles that govern the folding of protein chains, Les Prix Nobel en 1(1973) (1972) 103-119.

[34] D. Dunaway-Mariano, Enzyme function discovery, Structure 16 (2008) 1599-1600. https://doi.org/10.1016/j.str.2008.10.001

[35] G.A. Petsko, D. Ringe, Protein structure and function, New Science Press2004.

[36] L.H. Chen, G. Kenyon, F. Curtin, S. Harayama, M. Bembenek, G. Hajipour, C. Whitman, 4-Oxalocrotonate tautomerase, an enzyme composed of 62 amino acid residues per monomer, Journal of Biological Chemistry 267 (1992) 17716-17721. https://doi.org/10.1016/S0021-9258(19)37101-7

[37] S. Smith, The animal fatty acid synthase: one gene, one polypeptide, seven enzymes, The FASEB journal 8 (1994) 1248-1259. https://doi.org/10.1096/fasebj.8.15.8001737

[38] A.J.M. Ribeiro, G.L. Holliday, N. Furnham, J.D. Tyzack, K. Ferris, J.M. Thornton, Mechanism and Catalytic Site Atlas (M-CSA): a database of enzyme reaction

mechanisms and active sites, Nucleic acids research 46 (2018) D618-D623. https://doi.org/10.1093/nar/gkx1012

[39] H. Suzuki, Chapter 7: Active Site Structure, How Enzymes Work: From Structure to Function. Boca Raton (2015) 117-140. https://doi.org/10.1201/b18087-8

[40] G. Krauss, S. Benjamin, M. Lake, Biochemistry of signal transduction and regulation, Wiley Online Library2003. https://doi.org/10.1002/3527601864

[41] A.T. Brown, The Effect of Feed Additives on Male Broiler Performance, Mississippi State University, 2019.

[42] M. De Bolster, Glossary of terms used in bioinorganic chemistry (IUPAC recommendations 1997), Pure and applied chemistry 69 (1997) 1251-1304. https://doi.org/10.1351/pac199769061251

[43] D. Voet, J.G. Voet, C.W. Pratt, Fundamentals of biochemistry: life at the molecular level, John Wiley & Sons2016.

[44] A. Chapman-Smith, J.E. Cronan Jr, The enzymatic biotinylation of proteins: a post-translational modification of exceptional specificity, Trends in biochemical sciences 24 (1999) 359-363. https://doi.org/10.1016/S0968-0004(99)01438-3

[45] Z. Fisher, J.A. Hernandez Prada, C. Tu, D. Duda, C. Yoshioka, H. An, L. Govindasamy, D.N. Silverman, R. McKenna, Structural and kinetic characterization of active-site histidine as a proton shuttle in catalysis by human carbonic anhydrase II, Biochemistry 44 (2005) 1097-1105. https://doi.org/10.1021/bi0480279

[46] M.J. kadhim AL-Imam, B.A.L. AL-Rubaii, The influence of some amino acids, vitamins and anti-inflammatory drugs on activity of chondroitinase produced by Proteus vulgaris caused urinary tract infection, Iraqi Journal of Science 57 (2016) 2412-2421.

[47] P. Sarmah, D. Mahanta, Computational Methods for Enzyme Design and Its Biological Significance.

[48] T. Braunschweig, BRENDA The Comprehensive Enzyme Information System, 2014.

[49] D. Schomburg, I. Schomburg, S. Placzek, BRENDA-the Comprehensive enzyme information system, URL http://www. brenda-enzymes. org/enzyme. php (2015).

[50] S. Törnroth-Horsefield, R. Neutze, Opening and closing the metabolite gate, Proceedings of the National Academy of Sciences 105 (2008) 19565-19566. https://doi.org/10.1073/pnas.0810654106

[51] A. Cornish-Bowden, Why is uncompetitive inhibition so rare?: A possible explanation, with implications for the design of drugs and pesticides, FEBS letters 203 (1986) 3-6. https://doi.org/10.1016/0014-5793(86)81424-7

[52] R. Kopelmann, Fractal kinetics, Science (Washington DC) 241 (1988) 1620-1625. https://doi.org/10.1126/science.241.4873.1620

[53] J.M. Strelow, A perspective on the kinetics of covalent and irreversible inhibition, SLAS DISCOVERY: Advancing Life Sciences R&D 22 (2017) 3-20. https://doi.org/10.1177/1087057116671509

[54] N.C. Price, What is meant by 'competitive inhibition'?, Trends in biochemical sciences 4 (1979) N272-N273. https://doi.org/10.1016/0968-0004(79)90205-6

[55] D.S. Goodsell, The molecular perspective: methotrexate, The Oncologist 4 (1999) 340-341. https://doi.org/10.1634/theoncologist.4-4-340

[56] P. Wu, M.H. Clausen, T.E. Nielsen, Allosteric small-molecule kinase inhibitors, Pharmacology & therapeutics 156 (2015) 59-68. https://doi.org/10.1016/j.pharmthera.2015.10.002

[57] R. Seetharaman, M. Advani, S. Mali, S. Pawar, Enzymes as targets of Drug Action: an Overview, International Journal 3 (2020) 114.

[58] A.R. Jalalvand, Chemometrics in investigation of small molecule-biomacromolecule interactions: A review, International Journal of Biological Macromolecules (2021). https://doi.org/10.1016/j.ijbiomac.2021.03.184

[59] J.F. Fisher, S.O. Meroueh, S. Mobashery, Bacterial resistance to β-lactam antibiotics: compelling opportunism, compelling opportunity, Chemical reviews 105 (2005) 395-424. https://doi.org/10.1021/cr030102i

[60] D.S. Johnson, E. Weerapana, B.F. Cravatt, Strategies for discovering and derisking covalent, irreversible enzyme inhibitors, Future medicinal chemistry 2 (2010) 949-964. https://doi.org/10.4155/fmc.10.21

[61] I.V. Shevelev, U. Hübscher, The 3'-5' exonucleases, Nature reviews Molecular cell biology 3(5) (2002) 364-376. https://doi.org/10.1038/nrm804

[62] M. Ibba, D. Söll, Aminoacyl-tRNA synthesis, Annual review of biochemistry 69 (2000) 617-650. https://doi.org/10.1146/annurev.biochem.69.1.617

[63] N. Zenkin, Y. Yuzenkova, K. Severinov, Transcript-assisted transcriptional proofreading, Science 313 (2006) 518-520. https://doi.org/10.1126/science.1127422

[64] M. Yarus, S. Cline, L. Raftery, P. Wier, D. Bradley, The translational efficiency of tRNA is a property of the anticodon arm, Journal of Biological Chemistry 261 (1986) 10496-10505. https://doi.org/10.1016/S0021-9258(18)67412-5

[65] O.K. Tawfik, D. S, Enzyme promiscuity: a mechanistic and evolutionary perspective, Annual review of biochemistry 79 (2010) 471-505. https://doi.org/10.1146/annurev-biochem-030409-143718

[66] R.E. Ricklefs, Estimating diversification rates from phylogenetic information, Trends in ecology & evolution 22 (2007) 601-610. https://doi.org/10.1016/j.tree.2007.06.013

[67] E. Fischer, Effects of configuration on enzyme activity, Ber Dtsch Chem Ges 27 (1894) 2985-2993. https://doi.org/10.1002/cber.18940270364

[68] G.M. Cooper, The central role of enzymes as biological catalysts, Sinauer Associates2000.

[69] W.L. Jorgensen, Rusting of the lock and key model for protein-ligand binding, Science 254 (1991) 954-956. https://doi.org/10.1126/science.1719636

[70] C.S. Rye, S.G. Withers, Glycosidase mechanisms, Current opinion in chemical biology 4 (2000) 573-580. https://doi.org/10.1016/S1367-5931(00)00135-6

[71] R. Boyer, Chapter 6: Enzymes I, Reactions, Kinetics, and Inhibition, Concepts in Biochemistry (2002) 137-8.

[72] Y. Savir, T. Tlusty, Conformational proofreading: the impact of conformational changes on the specificity of molecular recognition, PloS one 2 (2007) e468. https://doi.org/10.1371/journal.pone.0000468

[73] K. Borgaonkar, R. Patil, RNA as ENZYMES and comparison of its properties with PROTEINS as ENZYMES, (2020). https://doi.org/10.26611/10021314

[74] A. Warshel, P.K. Sharma, M. Kato, Y. Xiang, H. Liu, M.H. Olsson, Electrostatic basis for enzyme catalysis, Chemical reviews 106 (2006) 3210-3235. https://doi.org/10.1021/cr0503106

[75] A.L. Lehninger, D.L. Nelson, M.M. Cox, Lehninger principles of biochemistry, New York: WH Freeman, 2013.

[76] S.J. Benkovic, S. Hammes-Schiffer, A perspective on enzyme catalysis, Science 301 (2003) 1196-1202. https://doi.org/10.1126/science.1085515

[77] W.P. Jencks, Catalysis in chemistry and enzymology, Courier Corporation1987.

[78] J. Villa, M. Štrajbl, T. Glennon, Y. Sham, Z. Chu, A. Warshel, How important are entropic contributions to enzyme catalysis?, Proceedings of the National Academy of Sciences 97 (2000) 11899-11904. https://doi.org/10.1073/pnas.97.22.11899

[79] A. Ramanathan, A. Savol, V. Burger, C.S. Chennubhotla, P.K. Agarwal, Protein conformational populations and functionally relevant substates, Accounts of chemical research 47 (2014) 149-156. https://doi.org/10.1021/ar400084s

[80] C.-J. Tsai, A. Del Sol, R. Nussinov, Protein allostery, signal transmission and dynamics: a classification scheme of allosteric mechanisms, Molecular Biosystems 5 (2009) 207-216. https://doi.org/10.1039/b819720b

[81] J.-P. Changeux, S.J. Edelstein, Allosteric mechanisms of signal transduction, Science 308 (2005) 1424-1428. https://doi.org/10.1126/science.1108595

[82] T. Hunter, A thousand and one protein kinases, Cell 50 (1987) 823-829. https://doi.org/10.1016/0092-8674(87)90509-5

[83] J. Berg, Powell BC, and Cheney RE, A millennial myosin census. Mol Biol Cell 12 (2001) 780-794. https://doi.org/10.1091/mbc.12.4.780

[84] E.A. Meighen, Molecular biology of bacterial bioluminescence, Microbiological reviews 55 (1991) 123-142. https://doi.org/10.1128/mr.55.1.123-142.1991

[85] E. De Clercq, Highlights in the development of new antiviral agents, Mini Rev Med Chem 2 (2002) 163-75. https://doi.org/10.2174/1389557024605474

[86] R.I. Mackie, B.A. White, Recent advances in rumen microbial ecology and metabolism: potential impact on nutrient output, Journal of dairy science 73 (1990) 2971-2995. https://doi.org/10.3168/jds.S0022-0302(90)78986-2

[87] R.G. Kurumbail, J.R. Kiefer, L.J. Marnett, Cyclooxygenase enzymes: catalysis and inhibition, Current opinion in structural biology 11 (2001) 752-760. https://doi.org/10.1016/S0959-440X(01)00277-9

[88] B.W. Doble, J.R. Woodgett, GSK-3: tricks of the trade for a multi-tasking kinase, Journal of cell science 116 (2003) 1175-1186. https://doi.org/10.1242/jcs.00384

[89] P. Bennett, I. Chopra, Molecular basis of beta-lactamase induction in bacteria, Antimicrobial Agents and Chemotherapy 37 (1993) 153. https://doi.org/10.1128/AAC.37.2.153

[90] G.G. Gibson, P. Skett, Induction and inhibition of drug metabolism, Introduction to drug metabolism, Springer1996, pp. 77-106. https://doi.org/10.1007/978-1-4899-6844-9_3

[91] G.M. Cohen, Caspases: the executioners of apoptosis, Biochemical Journal 326 (1997) 1-16. https://doi.org/10.1042/bj3260001

[92] H. Suzuki, Chapter 4: Effect of pH, Temperature, and High Pressure on Enzymatic Activity, How Enzymes Work: From Structure to Function (2015) 53-74. https://doi.org/10.1201/b18087-5

[93] C. Noree, B.K. Sato, R.M. Broyer, J.E. Wilhelm, Identification of novel filament-forming proteins in Saccharomyces cerevisiae and Drosophila melanogaster, Journal of Cell Biology 190 (2010) 541-551. https://doi.org/10.1083/jcb.201003001

[94] G.N. Aughey, J.-L. Liu, Metabolic regulation via enzyme filamentation, Critical reviews in biochemistry and molecular biology 51 (2016) 282-293. https://doi.org/10.3109/10409238.2016.1172555

[95] T. Nishimura, T. Iino, WO 2003080585 A1;(b) K. Kamata, M. Mitsuya, T. Nishimura, J. Eiki and Y. Nagata, Structure 12 (2004) 429. https://doi.org/10.1016/j.str.2004.02.005

[96] P. Froguel, H. Zouali, N. Vionnet, G. Velho, M. Vaxillaire, F. Sun, S. Lesage, M. Stoffel, J. Takeda, P. Passa, Familial hyperglycemia due to mutations in glucokinase--definition of a subtype of diabetes mellitus, New England Journal of Medicine 328 (1993) 697-702. https://doi.org/10.1056/NEJM199303113281005

[97] V. Renugopalakrishnan, R. Garduno-Juarez, G. Narasimhan, C. Verma, X. Wei, P. Li, Rational design of thermally stable proteins: relevance to bionanotechnology, Journal of nanoscience and nanotechnology 5 (2005) 1759-1767. https://doi.org/10.1166/jnn.2005.441

[98] K. Hult, P. Berglund, Engineered enzymes for improved organic synthesis, Current opinion in biotechnology 14 (2003) 395-400. https://doi.org/10.1016/S0958-1669(03)00095-8

[99] L. Jiang, E.A. Althoff, F.R. Clemente, L. Doyle, D. Röthlisberger, A. Zanghellini, J.L. Gallaher, J.L. Betker, F. Tanaka, C.F. Barbas, De novo computational design of retro-aldol enzymes, science 319 (2008) 1387-1391. https://doi.org/10.1126/science.1152692

[100] Y. Sun, J. Cheng, Hydrolysis of lignocellulosic materials for ethanol production: a review, Bioresource technology 83 (2002) 1-11. https://doi.org/10.1016/S0960-8524(01)00212-7

[101] O. Kirk, T., Vedel Borchert, and C. Crone Fuglsang. 2002. Industrial enzyme applications, Curr. Opin. Biotechnol 13 345-351. https://doi.org/10.1016/S0958-1669(02)00328-2

[102] C. Dulieu, M. Moll, J. Boudrant, D. Poncelet, Improved Performances and Control of Beer Fermentation Using Encapsulated α-Acetolactate Decarboxylase and Modeling, Biotechnology progress 16 (2000) 958-965. https://doi.org/10.1021/bp000128k

[103] R. Tarté, Ingredients in meat products: properties, functionality and applications, Springer2009. https://doi.org/10.1007/978-0-387-71327-4

[104] V.B. Khyade, A.J. Jeffreys, Biogeometric Model for the Magnification of the Mechanism of Enzyme Catalyzed Reaction through the Utilization of Principles of Nine-Point Circle for the Triangular Form of Lineweaver-Burk Plot in Biochemistry, International Journal of Biochemistry and Biomolecules 6 (2020) 1-23. https://doi.org/10.37121/ijesr.vol1.120

[105] S. Cakmakci, E. Dagdemir, A.A. Hayaloglu, M. Gurses, B. Cetin, D. Tahmas-Kahyaoglu, Effect of Penicillium roqueforti and incorporation of whey cheese on volatile profiles and sensory characteristics of mould-ripened Civil cheese, International Journal of Dairy Technology 66 (2013) 512-526. https://doi.org/10.1111/1471-0307.12042

[106] H. Guzmán-Maldonado, O. Paredes-López, C.G. Biliaderis, Amylolytic enzymes and products derived from starch: a review, Critical Reviews in Food Science & Nutrition 35 (1995) 373-403. https://doi.org/10.1080/10408399509527706

[107] I. Alkorta, C. Garbisu, M.J. Llama, J.L. Serra, Industrial applications of pectic enzymes: a review, Process Biochemistry 33 (1998) 21-28. https://doi.org/10.1016/S0032-9592(97)00046-0

[108] P. Bajpai, Application of enzymes in the pulp and paper industry, Biotechnology progress 15 (1999) 147-157. https://doi.org/10.1021/bp990013k

[109] C. Begley, S. Paragina, A. Sporn, An analysis of contact lens enzyme cleaners, Journal of the American Optometric Association 61 (1990) 190-194.

[110] P.L. Farris, Economic Growth and Organization of the US Corn Starch Industry, Starch, Elsevier2009, pp. 11-21. https://doi.org/10.1016/B978-0-12-746275-2.00002-1

[111] D. Brady, J. Jordaan, Advances in enzyme immobilisation, Biotechnology letters 31 (2009) 1639-1650. https://doi.org/10.1007/s10529-009-0076-4

[112] C.K. Lee, A.N. Au-Duong, Enzyme immobilization on nanoparticles: recent applications, Emerging Areas in Bioengineering 1 (2018) 67-80. https://doi.org/10.1002/9783527803293.ch4

[113] C. Spahn, S.D. Minteer, Enzyme immobilization in biotechnology, Recent patents on engineering 2 (2008) 195-200. https://doi.org/10.2174/187221208786306333

[114] R. Johnson, Z.-G. Wang, F. Arnold, Surface site heterogeneity and lateral interactions in multipoint protein adsorption, The Journal of Physical Chemistry 100 (1996) 5134-5139. https://doi.org/10.1021/jp9523682

[115] S.V. Rao, K.W. Anderson, L.G. Bachas, Oriented immobilization of proteins, Microchimica Acta 128 (1998) 127-143. https://doi.org/10.1007/BF01243043

[116] J. Fu, J. Reinhold, N.W. Woodbury, Peptide-modified surfaces for enzyme immobilization, PLoS One 6 (2011) e18692. https://doi.org/10.1371/journal.pone.0018692

[117] V. Singh, M. Sardar, M.N. Gupta, Immobilization of enzymes by bioaffinity layering, Immobilization of Enzymes and Cells, Springer2013, pp. 129-137. https://doi.org/10.1007/978-1-62703-550-7_9

[118] M. Hartmann, X. Kostrov, Immobilization of enzymes on porous silicas-benefits and challenges, Chemical Society Reviews 42 (2013) 6277-6289. https://doi.org/10.1039/c3cs60021a

[119] D. Li, Q. He, Y. Cui, L. Duan, J. Li, Immobilization of glucose oxidase onto gold nanoparticles with enhanced thermostability, Biochemical and biophysical research communications 355 (2007) 488-493. https://doi.org/10.1016/j.bbrc.2007.01.183

[120] C.-Y. Yu, L.-Y. Huang, I. Kuan, S.-L. Lee, Optimized production of biodiesel from waste cooking oil by lipase immobilized on magnetic nanoparticles, International journal of molecular sciences 14 (2013) 24074-24086. https://doi.org/10.3390/ijms141224074

[121] J. Hou, G. Dong, B. Xiao, C. Malassigne, V. Chen, Preparation of titania based biocatalytic nanoparticles and membranes for CO 2 conversion, Journal of Materials Chemistry A 3 (2015) 3332-3342. https://doi.org/10.1039/C4TA05760K

[122] S. Wang, P. Su, J. Huang, J. Wu, Y. Yang, Magnetic nanoparticles coated with immobilized alkaline phosphatase for enzymolysis and enzyme inhibition assays, Journal of Materials Chemistry B 1 (2013) 1749-1754. https://doi.org/10.1039/c3tb00562c

[123] V. Hooda, Immobilization and kinetics of catalase on calcium carbonate nanoparticles attached epoxy support, Applied biochemistry and biotechnology 172 (2014) 115-130. https://doi.org/10.1007/s12010-013-0498-2

[124] J.T. Holland, C. Lau, S. Brozik, P. Atanassov, S. Banta, Engineering of glucose oxidase for direct electron transfer via site-specific gold nanoparticle conjugation, Journal of the American Chemical Society 133 (2011) 19262-19265. https://doi.org/10.1021/ja2071237

[125] S. Zhang, N. Wang, H. Yu, Y. Niu, C. Sun, Covalent attachment of glucose oxidase to an Au electrode modified with gold nanoparticles for use as glucose biosensor, Bioelectrochemistry 67 (2005) 15-22. https://doi.org/10.1016/j.bioelechem.2004.12.002

[126] S. Datta, L.R. Christena, Y.R.S. Rajaram, Enzyme immobilization: an overview on techniques and support materials, 3 Biotech 3 (2013) 1-9. https://doi.org/10.1007/s13205-012-0071-7

[127] A. Deshpande, S. D'souza, G. Nadkarni, Coimmobilization of D-amino acid oxidase and catalase by entrapment ofTrigonopsis variabilis in radiation polymerised Polyacrylamide beads, Journal of Biosciences 11 (1987) 137-144. https://doi.org/10.1007/BF02704664

[128] Z. Grosová, M. Rosenberg, M. Rebroš, M. Šipocz, B. Sedláčková, Entrapment of β-galactosidase in polyvinylalcohol hydrogel, Biotechnology Letters 30 (2008) 763-767. https://doi.org/10.1007/s10529-007-9606-0

[129] R. Sheldon, Cross-linked enzyme aggregates (CLEA® s): stable and recyclable biocatalysts, Biochemical Society Transactions 35 (2007) 1583-1587. https://doi.org/10.1042/BST0351583

[130] R.A. Sheldon, Enzyme immobilization: the quest for optimum performance, Advanced Synthesis & Catalysis 349 (2007) 1289-1307. https://doi.org/10.1002/adsc.200700082

[131] A.A. Homaei, R. Sariri, F. Vianello, R. Stevanato, Enzyme immobilization: an update, Journal of chemical biology 6 (2013) 185-205. https://doi.org/10.1007/s12154-013-0102-9

[132] B. Katzbauer, M. Narodoslawsky, A. Moser, Classification system for immobilization techniques, Bioprocess Engineering 12 (1995) 173-179. https://doi.org/10.1007/BF01767463

[133] C.A. White, J.F. Kennedy, Popular matrices for enzyme and other immobilizations, Enzyme and Microbial Technology 2 (1980) 82-90. https://doi.org/10.1016/0141-0229(80)90062-9

[134] S. Sommaruga, E. Galbiati, J. Peñaranda-Avila, C. Brambilla, P. Tortora, M. Colombo, D. Prosperi, Immobilization of carboxypeptidase from Sulfolobus solfataricus on magnetic nanoparticles improves enzyme stability and functionality in organic media, BMC biotechnology 14 (2014) 1-9. https://doi.org/10.1186/1472-6750-14-82

[135] E.D. Hood, M. Chorny, C.F. Greineder, I.S. Alferiev, R.J. Levy, V.R. Muzykantov, Endothelial targeting of nanocarriers loaded with antioxidant enzymes for protection against vascular oxidative stress and inflammation, Biomaterials 35 (2014) 3708-3715. https://doi.org/10.1016/j.biomaterials.2014.01.023

[136] C.-C. Yu, Y.-Y. Kuo, C.-F. Liang, W.-T. Chien, H.-T. Wu, T.-C. Chang, F.-D. Jan, C.-C. Lin, Site-specific immobilization of enzymes on magnetic nanoparticles and their use in organic synthesis, Bioconjugate chemistry 23 (2012) 714-724. https://doi.org/10.1021/bc200396r

[137] L. Zhou, J. Wu, H. Zhang, Y. Kang, J. Guo, C. Zhang, J. Yuan, X. Xing, Magnetic nanoparticles for the affinity adsorption of maltose binding protein (MBP) fusion enzymes, Journal of Materials Chemistry 22 (2012) 6813-6818. https://doi.org/10.1039/c2jm16778f

[138] E. Jang, S. Park, S. Park, Y. Lee, D.N. Kim, B. Kim, W.G. Koh, Fabrication of poly (ethylene glycol)-based hydrogels entrapping enzyme-immobilized silica nanoparticles, Polymers for Advanced Technologies 21 (2010) 476-482. https://doi.org/10.1002/pat.1455

[139] S. Wang, E.S. Humphreys, S.-Y. Chung, D.F. Delduco, S.R. Lustig, H. Wang, K.N. Parker, N.W. Rizzo, S. Subramoney, Y.-M. Chiang, Peptides with selective affinity for carbon nanotubes, Nature materials 2 (2003) 196-200. https://doi.org/10.1038/nmat833

[140] P. Ji, H. Tan, X. Xu, W. Feng, Lipase covalently attached to multiwalled carbon nanotubes as an efficient catalyst in organic solvent, AIChE journal 56 (2010) 3005-3011. https://doi.org/10.1002/aic.12180

[141] H. Tan, W. Feng, P. Ji, Lipase immobilized on magnetic multi-walled carbon nanotubes, Bioresource technology 115 (2012) 172-176. https://doi.org/10.1016/j.biortech.2011.10.066

[142] W. Feng, X. Sun, P. Ji, Activation mechanism of Yarrowia lipolytica lipase immobilized on carbon nanotubes, Soft Matter 8 (2012) 7143-7150. https://doi.org/10.1039/c2sm25231g

[143] A.K. Geim, K.S. Novoselov, The rise of graphene, Nanoscience and technology: a collection of reviews from nature journals, World Scientific2010, pp. 11-19. https://doi.org/10.1142/9789814287005_0002

[144] O.C. Compton, S.T. Nguyen, Graphene oxide, highly reduced graphene oxide, and graphene: versatile building blocks for carbon-based materials, small 6 (2010) 711-723. https://doi.org/10.1002/smll.200901934

[145] S. Hermanová, M. Zarevúcká, D. Bouša, M. Pumera, Z. Sofer, Graphene oxide immobilized enzymes show high thermal and solvent stability, Nanoscale 7 (2015) 5852-5858. https://doi.org/10.1039/C5NR00438A

[146] E.C. Dreaden, A.M. Alkilany, X. Huang, C.J. Murphy, M.A. El-Sayed, The golden age: gold nanoparticles for biomedicine, Chemical Society Reviews 41 (2012) 2740-2779. https://doi.org/10.1039/C1CS15237H

[147] I. Venditti, C. Palocci, L. Chronopoulou, I. Fratoddi, L. Fontana, M. Diociaiuti, M.V. Russo, Candida rugosa lipase immobilization on hydrophilic charged gold nanoparticles as promising biocatalysts: Activity and stability investigations, Colloids and Surfaces B: Biointerfaces 131 (2015) 93-101. https://doi.org/10.1016/j.colsurfb.2015.04.046

[148] B.J. Sanghavi, S.M. Mobin, P. Mathur, G.K. Lahiri, A.K. Srivastava, Biomimetic sensor for certain catecholamines employing copper (II) complex and silver nanoparticle modified glassy carbon paste electrode, Biosensors and Bioelectronics 39 (2013) 124-132. https://doi.org/10.1016/j.bios.2012.07.008

[149] M. Falasconi, M. Pardo, G. Sberveglieri, I. Riccò, A. Bresciani, The novel EOS835 electronic nose and data analysis for evaluating coffee ripening, Sensors and Actuators B: Chemical 110 (2005) 73-80. https://doi.org/10.1016/j.snb.2005.01.019

[150] N.J. Vickers, Animal communication: when i'm calling you, will you answer too?, Current biology 27 (2017) R713-R715. https://doi.org/10.1016/j.cub.2017.05.064

[151] J. Jeong, T.H. Ha, B.H. Chung, Enhanced reusability of hexa-arginine-tagged esterase immobilized on gold-coated magnetic nanoparticles, Analytica Chimica Acta 569 (2006) 203-209. https://doi.org/10.1016/j.aca.2006.03.102

[152] A.J. Wikieł, I. Datsenko, M. Vera, W. Sand, Impact of Desulfovibrio alaskensis biofilms on corrosion behaviour of carbon steel in marine environment,

Bioelectrochemistry 97 (2014) 52-60.
https://doi.org/10.1016/j.bioelechem.2013.09.008

[153] C. Radhakumary, K. Sreenivasan, Naked eye detection of glucose in urine using glucose oxidase immobilized gold nanoparticles, Analytical chemistry 83 (2011) 2829-2833. https://doi.org/10.1021/ac1032879

[154] U. Saxena, M. Chakraborty, P. Goswami, Covalent immobilization of cholesterol oxidase on self-assembled gold nanoparticles for highly sensitive amperometric detection of cholesterol in real samples, Biosensors and Bioelectronics 26 (2011) 3037-3043. https://doi.org/10.1016/j.bios.2010.12.009

[155] X. Ren, D. Chen, X. Meng, F. Tang, A. Du, L. Zhang, Amperometric glucose biosensor based on a gold nanorods/cellulose acetate composite film as immobilization matrix, Colloids and Surfaces B: Biointerfaces 72 (2009) 188-192. https://doi.org/10.1016/j.colsurfb.2009.04.003

[156] A. Homaei, D. Saberi, Immobilization of α-amylase on gold nanorods: An ideal system for starch processing, Process Biochemistry 50 (2015) 1394-1399. https://doi.org/10.1016/j.procbio.2015.06.002

[157] S. Aravamudhan, N.S. Ramgir, S. Bhansali, Electrochemical biosensor for targeted detection in blood using aligned Au nanowires, Sensors and Actuators B: Chemical 127 (2007) 29-35. https://doi.org/10.1016/j.snb.2007.07.008

[158] T. Xu, J. Zhang, H. Chi, F. Cao, Multifunctional properties of organic-inorganic hybrid nanocomposites based on chitosan derivatives and layered double hydroxides for ocular drug delivery, Acta biomaterialia 36 (2016) 152-163. https://doi.org/10.1016/j.actbio.2016.02.041

[159] S. He, Z. An, M. Wei, D.G. Evans, X. Duan, Layered double hydroxide-based catalysts: nanostructure design and catalytic performance, Chemical communications 49 (2013) 5912-5920. https://doi.org/10.1039/c3cc42137f

[160] J. Wang, P. Wang, H. Wang, J. Dong, W. Chen, X. Wang, S. Wang, T. Hayat, A. Alsaedi, X. Wang, Preparation of molybdenum disulfide coated Mg/Al layered double hydroxide composites for efficient removal of chromium (VI), ACS Sustainable Chemistry & Engineering 5 (2017) 7165-7174. https://doi.org/10.1021/acssuschemeng.7b01347

[161] C. Mousty, V. Prévot, Hybrid and biohybrid layered double hydroxides for electrochemical analysis, Analytical and bioanalytical chemistry 405 (2013) 3513-3523. https://doi.org/10.1007/s00216-013-6797-1

[162] P.A. Harris, R. Taylor, B.L. Minor, V. Elliott, M. Fernandez, L. O'Neal, L. McLeod, G. Delacqua, F. Delacqua, J. Kirby, The REDCap consortium: Building an international community of software platform partners, Journal of biomedical informatics 95 (2019) 103208. https://doi.org/10.1016/j.jbi.2019.103208

[163] M. Foresti, G. Valle, R. Bonetto, M. Ferreira, L. Briand, FTIR, SEM and fractal dimension characterization of lipase B from Candida antarctica immobilized onto titania at selected conditions, Applied Surface Science 256 (2010) 1624-1635. https://doi.org/10.1016/j.apsusc.2009.09.083

[164] R.P. Lopes, R.C. Reyes, R. Romero-González, A.G. Frenich, J.L.M. Vidal, Development and validation of a multiclass method for the determination of veterinary drug residues in chicken by ultra high performance liquid chromatography-tandem mass spectrometry, Talanta 89 (2012) 201-208. https://doi.org/10.1016/j.talanta.2011.11.082

[165] W. Xie, N. Ma, Immobilized lipase on Fe3O4 nanoparticles as biocatalyst for biodiesel production, Energy & Fuels 23 (2009) 1347-1353. https://doi.org/10.1021/ef800648y

[166] S.M.U. Ali, O. Nur, M. Willander, B. Danielsson, A fast and sensitive potentiometric glucose microsensor based on glucose oxidase coated ZnO nanowires grown on a thin silver wire, Sensors and Actuators B: Chemical 145 (2010) 869-874. https://doi.org/10.1016/j.snb.2009.12.072

[167] D.-M. Liu, J. Chen, Y.-P. Shi, Advances on methods and easy separated support materials for enzymes immobilization, TrAC Trends in Analytical Chemistry 102 (2018) 332-342. https://doi.org/10.1016/j.trac.2018.03.011

[168] H. Vaghari, H. Jafarizadeh-Malmiri, M. Mohammadlou, A. Berenjian, N. Anarjan, N. Jafari, S. Nasiri, Application of magnetic nanoparticles in smart enzyme immobilization, Biotechnology letters 38 (2016) 223-233. https://doi.org/10.1007/s10529-015-1977-z

[169] R. Marcus, Sutin, N. flioc/iim, Biophys. Acta 811 (1985) 265. https://doi.org/10.1016/0304-4173(85)90014-X

[170] Y. Zhao, X. Li, Y. Yang, S. Si, C. Deng, H. Wu, A simple aptasensor for Aβ40 oligomers based on tunable mismatched base pairs of dsDNA and graphene oxide, Biosensors and Bioelectronics 149 (2020) 111840. https://doi.org/10.1016/j.bios.2019.111840

[171] M.A. Rahman, N.-H. Kwon, M.-S. Won, E.S. Choe, Y.-B. Shim, Functionalized conducting polymer as an enzyme-immobilizing substrate: an amperometric glutamate microbiosensor for in vivo measurements, Analytical Chemistry 77 (2005) 4854-4860. https://doi.org/10.1021/ac050558v

[172] L. Kergoat, B. Piro, D.T. Simon, M.C. Pham, V. Noël, M. Berggren, Detection of glutamate and acetylcholine with organic electrochemical transistors based on conducting polymer/platinum nanoparticle composites, Advanced Materials 26 (2014) 5658-5664. https://doi.org/10.1002/adma.201401608

[173] A.A. Abdelwahab, W.C.A. Koh, H.-B. Noh, Y.-B. Shim, A selective nitric oxide nanocomposite biosensor based on direct electron transfer of microperoxidase: removal of interferences by co-immobilized enzymes, Biosensors and Bioelectronics 26 (2010) 1080-1086. https://doi.org/10.1016/j.bios.2010.08.070

[174] Z.-X. Tang, J.-Q. Qian, L.-E. Shi, Preparation of chitosan nanoparticles as carrier for immobilized enzyme, Applied Biochemistry and Biotechnology 136 (2007) 77-96. https://doi.org/10.1007/BF02685940

[175] M. Babaki, M. Yousefi, Z. Habibi, M. Mohammadi, P. Yousefi, J. Mohammadi, J. Brask, Enzymatic production of biodiesel using lipases immobilized on silica nanoparticles as highly reusable biocatalysts: effect of water, t-butanol and blue silica gel contents, Renewable Energy 91 (2016) 196-206. https://doi.org/10.1016/j.renene.2016.01.053

[176] X.-J. Huang, P.-C. Chen, F. Huang, Y. Ou, M.-R. Chen, Z.-K. Xu, Immobilization of Candida rugosa lipase on electrospun cellulose nanofiber membrane, Journal of Molecular Catalysis B: Enzymatic 70 (2011) 95-100. https://doi.org/10.1016/j.molcatb.2011.02.010

[177] W. Li, P. Zheng, J. Guo, J. Ji, M. Zhang, Z. Zhang, E. Zhan, G. Abbas, Characteristics of self-alkalization in high-rate denitrifying automatic circulation (DAC) reactor fed with methanol and sodium acetate, Bioresource technology 154 (2014) 44-50. https://doi.org/10.1016/j.biortech.2013.11.097

[178] N. An, C.H. Zhou, X.Y. Zhuang, D.S. Tong, W.H. Yu, Immobilization of enzymes on clay minerals for biocatalysts and biosensors, Applied Clay Science 114 (2015) 283-296. https://doi.org/10.1016/j.clay.2015.05.029

[179] X. Wu, J. Ge, C. Yang, M. Hou, Z. Liu, Facile synthesis of multiple enzyme-containing metal-organic frameworks in a biomolecule-friendly environment, Chemical Communications 51 (2015) 13408-13411. https://doi.org/10.1039/C5CC05136C

[180] Z. Zhou, M. Hartmann, Progress in enzyme immobilization in ordered mesoporous materials and related applications, Chemical Society Reviews 42 (2013) 3894-3912. https://doi.org/10.1039/c3cs60059a

[181] X. Lian, Y. Fang, E. Joseph, Q. Wang, J. Li, S. Banerjee, C. Lollar, X. Wang, H.-C. Zhou, Enzyme-MOF (metal-organic framework) composites, Chemical Society Reviews 46 (2017) 3386-3401. https://doi.org/10.1039/C7CS00058H

[182] L.C. Clark Jr, C. Lyons, Electrode systems for continuous monitoring in cardiovascular surgery, Annals of the New York Academy of sciences 102 (1962) 29-45. https://doi.org/10.1111/j.1749-6632.1962.tb13623.x

[183] P. Bollella, L. Gorton, Enzyme based amperometric biosensors, Current Opinion in Electrochemistry 10 (2018) 157-173. https://doi.org/10.1016/j.coelec.2018.06.003

[184] S. Kurbanoglu, M.N. Zafar, F. Tasca, I. Aslam, O. Spadiut, D. Leech, D. Haltrich, L. Gorton, Amperometric Flow Injection Analysis of Glucose and Galactose Based on Engineered Pyranose 2-Oxidases and Osmium Polymers for Biosensor Applications, Electroanalysis 30 (2018) 1496-1504. https://doi.org/10.1002/elan.201800096

[185] D. Grieshaber, R. MacKenzie, J. Vörös, E. Reimhult, Electrochemical biosensors-sensor principles and architectures, Sensors 8 (2008) 1400-1458. https://doi.org/10.3390/s80314000

[186] T.T. Dung, Y. Oh, S.-J. Choi, I.-D. Kim, M.-K. Oh, M. Kim, Applications and advances in bioelectronic noses for odour sensing, Sensors 18 (2018) 103. https://doi.org/10.3390/s18010103

[187] C.D. Fung, P.W. Cheung, W.H. Ko, A generalized theory of an electrolyte-insulator-semiconductor field-effect transistor, IEEE Transactions on Electron Devices 33 (1986) 8-18. https://doi.org/10.1109/T-ED.1986.22429

[188] W.-Y. Lee, S.-R. Kim, T.-H. Kim, K.S. Lee, M.-C. Shin, J.-K. Park, Sol-gel-derived thick-film conductometric biosensor for urea determination in serum, Analytica Chimica Acta 404 (2000) 195-203. https://doi.org/10.1016/S0003-2670(99)00699-6

[189] S.R. Mikkelsen, G.A. Rechnitz, Conductometric tranducers for enzyme-based biosensors, Analytical chemistry 61 (1989) 1737-1742. https://doi.org/10.1021/ac00190a029

[190] U. Bilitewski, W. Drewes, R. Schmid, Thick film biosensors for urea, Sensors and Actuators B: Chemical 7 (1992) 321-326. https://doi.org/10.1016/0925-4005(92)80317-Q

[191] A. Shul'ga, A. Soldatkin, A. El'skaya, S. Dzyadevich, S. Patskovsky, V. Strikha, Thin-film conductometric biosensors for glucose and urea determination, Biosensors and Bioelectronics 9 (1994) 217-223. https://doi.org/10.1016/0956-5663(94)80124-X

[192] M. Castillo-Ortega, D. Rodriguez, J. Encinas, M. Plascencia, F. Mendez-Velarde, R. Olayo, Conductometric uric acid and urea biosensor prepared from electroconductive polyaniline-poly (n-butyl methacrylate) composites, Sensors and Actuators B: Chemical 85 (2002) 19-25. https://doi.org/10.1016/S0925-4005(02)00045-X

[193] W.O. Ho, S. Krause, C.J. McNeil, J.A. Pritchard, R.D. Armstrong, D. Athey, K. Rawson, Electrochemical sensor for measurement of urea and creatinine in serum based on ac impedance measurement of enzyme-catalyzed polymer transformation, Analytical Chemistry 71 (1999) 1940-1946. https://doi.org/10.1021/ac981367d

[194] S.V. Dzyadevych, J.-M. Chovelon, A comparative photodegradation studies of methyl parathion by using Lumistox test and conductometric biosensor technique, Materials Science and Engineering: C 21 (2002) 55-60. https://doi.org/10.1016/S0928-4931(02)00058-9

[195] S.V. Dzyadevych, A.P. Soldatkin, J.-M. Chovelon, Assessment of the toxicity of methyl parathion and its photodegradation products in water samples using conductometric enzyme biosensors, Analytica Chimica Acta 459 (2002) 33-41. https://doi.org/10.1016/S0003-2670(02)00083-1

[196] P. Jin, A. Yamaguchi, F.A. Oi, S. Matsuo, J. Tan, H. Misawa, Glucose sensing based on interdigitated array microelectrode, Analytical sciences 17 (2001) 841-846. https://doi.org/10.2116/analsci.17.841

[197] S.V. Dzyadevych, V.N. Arkhypova, Y.I. Korpan, V. Anna, A.P. Soldatkin, N. Jaffrezic-Renault, C. Martelet, Conductometric formaldehyde sensitive biosensor with specifically adapted analytical characteristics, Analytica chimica acta 445 (2001) 47-55. https://doi.org/10.1016/S0003-2670(01)01249-1

[198] M. Marrakchi, S.V. Dzyadevych, F. Lagarde, C. Martelet, N. Jaffrezic-Renault, Conductometric biosensor based on glucose oxidase and beta-galactosidase for specific lactose determination in milk, Materials Science and Engineering: C 28 (2008) 872-875. https://doi.org/10.1016/j.msec.2007.10.046

[199] T.M. Anh, S.V. Dzyadevych, M.C. Van, N.J. Renault, C.N. Duc, J.-M. Chovelon, Conductometric tyrosinase biosensor for the detection of diuron, atrazine and its main metabolites, Talanta 63 (2004) 365-370. https://doi.org/10.1016/j.talanta.2003.11.008

[200] S.K. Kirdeciler, E. Soy, S. Öztürk, I. Kucherenko, O. Soldatkin, S. Dzyadevych, B. Akata, A novel urea conductometric biosensor based on zeolite immobilized urease, Talanta 85 (2011) 1435-1441. https://doi.org/10.1016/j.talanta.2011.06.034

[201] A.N. Hendji, N. Jaffrezic-Renault, C. Martelet, A. Shul'ga, S. Dzydevich, A. Soldatkin, A. El'skaya, Enzyme biosensor based on a micromachined interdigitated conductometric transducer: application to the detection of urea, glucose, acetyl-andbutyrylcholine chlordes, Sensors and Actuators B: Chemical 21 (1994) 123-129. https://doi.org/10.1016/0925-4005(94)80013-8

[202] H.H. Nguyen, S.H. Lee, U.J. Lee, C.D. Fermin, M. Kim, Immobilized enzymes in biosensor applications, Materials 12 (2019) 121. https://doi.org/10.3390/ma12010121

[203] V. Arkhypova, S. Dzyadevych, A. Soldatkin, A. El'Skaya, N. Jaffrezic-Renault, H. Jaffrezic, C. Martelet, Multibiosensor based on enzyme inhibition analysis for determination of different toxic substances, Talanta 55 (2001) 919-927. https://doi.org/10.1016/S0039-9140(01)00495-7

[204] S.V. Dzyadevych, A.P. Soldatkin, V.N. Arkhypova, V. Anna, J.-M. Chovelon, C.A. Georgiou, C. Martelet, N. Jaffrezic-Renault, Early-warning electrochemical biosensor system for environmental monitoring based on enzyme inhibition, Sensors and Actuators B: Chemical 105 (2005) 81-87. https://doi.org/10.1016/S0925-4005(04)00115-7

[205] T. Sergeyeva, N. Lavrik, A. Rachkov, Z. Kazantseva, S. Piletsky, A. El'skaya, Hydrogen peroxide-sensitive enzyme sensor based on phthalocyanine thin film, Analytica Chimica Acta 391 (1999) 289-297. https://doi.org/10.1016/S0003-2670(99)00203-2

[206] D. Cullen, R. Sethi, C. Lowe, Multi-analyte miniature conductance biosensor, Analytica Chimica Acta 231 (1990) 33-40. https://doi.org/10.1016/S0003-2670(00)86394-1

[207] S. Myler, S.D. Collyer, F. Davis, D.D. Gornall, S.P. Higson, Sonochemically fabricated microelectrode arrays for biosensors: Part III. AC impedimetric study of aerobic and anaerobic response of alcohol oxidase within polyaniline, Biosensors and Bioelectronics 21 (2005) 666-671. https://doi.org/10.1016/j.bios.2004.12.012

[208] R.K. Shervedani, A.H. Mehrjardi, N. Zamiri, A novel method for glucose determination based on electrochemical impedance spectroscopy using glucose oxidase self-assembled biosensor, Bioelectrochemistry 69 (2006) 201-208. https://doi.org/10.1016/j.bioelechem.2006.01.003

[209] D. Zane, G. Appetecchi, C. Bianchini, S. Passerini, A. Curulli, An impedimetric glucose biosensor based on overoxidized polypyrrole thin film, Electroanalysis 23 (2011) 1134-1141. https://doi.org/10.1002/elan.201000576

[210] F. Abdelmalek, M. Shadaram, H. Boushriha, Ellipsometry measurements and impedance spectroscopy on Langmuir-Blodgett membranes on Si/SiO2 for ion sensitive sensor, Sensors and Actuators B: Chemical 72(3) (2001) 208-213. https://doi.org/10.1016/S0925-4005(00)00657-2

[211] N. Bouyahia, M.L. Hamlaoui, M. Hnaien, F. Lagarde, N. Jaffrezic-Renault, Impedance spectroscopy and conductometric biosensing for probing catalase reaction with cyanide as ligand and inhibitor, Bioelectrochemistry 80 (2011) 155-161. https://doi.org/10.1016/j.bioelechem.2010.07.006

[212] M. Cortina, M. Esplandiu, S. Alegret, M. Del Valle, Urea impedimetric biosensor based on polymer degradation onto interdigitated electrodes, Sensors and Actuators B: Chemical 118 (2006) 84-89. https://doi.org/10.1016/j.snb.2006.04.062

[213] N. Zehani, S.V. Dzyadevych, R. Kherrat, N.J. Jaffrezic-Renault, Sensitive impedimetric biosensor for direct detection of diazinon based on lipases, Frontiers in chemistry 2 (2014) 44. https://doi.org/10.3389/fchem.2014.00044

[214] M. Shamsipur, M. Asgari, M.G. Maragheh, A.A. Moosavi-Movahedi, A novel impedimetric nanobiosensor for low level determination of hydrogen peroxide based on biocatalysis of catalase, Bioelectrochemistry 83 (2012) 31-37. https://doi.org/10.1016/j.bioelechem.2011.08.003

[215] R. Maalouf, H. Chebib, Y. Saïkali, O. Vittori, M. Sigaud, N. Jaffrezic-Renault, Amperometric and impedimetric characterization of a glutamate biosensor based on Nafion® and a methyl viologen modified glassy carbon electrode, Biosensors and Bioelectronics 22 (2007) 2682-2688. https://doi.org/10.1016/j.bios.2006.11.003

[216] E.T. Hwang, M.B. Gu, Enzyme stabilization by nano/microsized hybrid materials, Engineering in Life Sciences 13 (2013) 49-61. https://doi.org/10.1002/elsc.201100225

[217] X. Wang, P. Dou, P. Zhao, C. Zhao, Y. Ding, P. Xu, Immobilization of lipases onto magnetic Fe3O4 nanoparticles for application in biodiesel production, ChemSusChem: Chemistry & Sustainability Energy & Materials 2 (2009) 947-950. https://doi.org/10.1002/cssc.200900174

[218] W. Xie, N. Ma, Enzymatic transesterification of soybean oil by using immobilized lipase on magnetic nano-particles, Biomass and Bioenergy 34 (2010) 890-896. https://doi.org/10.1016/j.biombioe.2010.01.034

[219] X. Wang, X. Liu, X. Yan, P. Zhao, Y. Ding, P. Xu, Enzyme-nanoporous gold biocomposite: excellent biocatalyst with improved biocatalytic performance and stability, PLoS One 6 (2011) e24207. https://doi.org/10.1371/journal.pone.0024207

[220] M.L. Verma, M. Puri, C.J. Barrow, Recent trends in nanomaterials immobilised enzymes for biofuel production, Critical reviews in biotechnology 36 (2016) 108-119. https://doi.org/10.3109/07388551.2014.928811

[221] M. Jacoby, The mystery of hot gold nanoparticles, Chemical & Engineering News 91 (2013) 44-45. https://doi.org/10.1021/cen-09113-scitech2

[222] S.-F. Li, Y.-H. Fan, R.-F. Hu, W.-T. Wu, Pseudomonas cepacia lipase immobilized onto the electrospun PAN nanofibrous membranes for biodiesel production from soybean oil, Journal of Molecular Catalysis B: Enzymatic 72 (2011) 40-45. https://doi.org/10.1016/j.molcatb.2011.04.022

[223] S. Gaikwad, A.P. Ingle, S.S. da Silva, M. Rai, Immobilized nanoparticles-mediated enzymatic hydrolysis of cellulose for clean sugar production: A novel approach, Current Nanoscience 15 (2019) 296-303. https://doi.org/10.2174/1573413714666180611081759

[224] A.N. Kozitsina, T.S. Svalova, N.N. Malysheva, A.V. Okhokhonin, M.B. Vidrevich, K.Z. Brainina, Sensors based on bio and biomimetic receptors in medical diagnostic, environment, and food analysis, Biosensors 8 (2018) 35. https://doi.org/10.3390/bios8020035

[225] N. Noah, Design and Synthesis of Nanostructured Materials forSensor Applications, (2020). https://doi.org/10.1155/2020/8855321

[226] N.M. Noah, Design and Synthesis of Nanostructured Materials for Sensor Applications, Journal of Nanomaterials 2020 (2020). https://doi.org/10.1155/2020/8855321

[227] M. Rai, A.P. Ingle, S. Gaikwad, K.J. Dussán, S.S. da Silva, Role of nanoparticles in enzymatic hydrolysis of lignocellulose in ethanol, Nanotechnology for bioenergy and biofuel production, Springer2017, pp. 153-171. https://doi.org/10.1007/978-3-319-45459-7_7

[228] S.K. Sahu, S. Sahu, V. Nourani, C.A. Board, U. Shanker, R. Shanker, V. Kumari, K.K. Bansal, D. Garg, V.A. Athavale, Untitled-International Journal of Engineering and Advanced.

[229] A.K. Shukla, M. Verma, A. Acharya, Biomolecules Immobilized Nanomaterials and Their Biological Applications, Nanomaterial-Based Biomedical Applications in

Molecular Imaging, Diagnostics and Therapy, Springer2020, pp. 79-101. https://doi.org/10.1007/978-981-15-4280-0_5

[230] M. Singhvi, B.S. Kim, Current Developments in Lignocellulosic Biomass Conversion into Biofuels Using Nanobiotechology Approach, Energies 13 (2020) 5300. https://doi.org/10.3390/en13205300

[231] G.S. Wilson, Y. Hu, Enzyme-based biosensors for in vivo measurements, Chemical reviews 100(7) (2000) 2693-2704. https://doi.org/10.1021/cr990003y

[232] C. Karunakaran, R. Rajkumar, K. Bhargava, Introduction to biosensors, Biosensors and bioelectronics, Elsevier2015, pp. 1-68. https://doi.org/10.1016/B978-0-12-803100-1.00001-3

[233] P. Sistani, L. Sofimaryo, Z.R. Masoudi, A. Sayad, R. Rahimzadeh, B. Salehi, A penicillin biosensor by using silver nanoparticles, Int. J. Electrochem. Sci 9 (2014) 6201-6212.

[234] S. Bourigua, A. Maaref, F. Bessueille, N.J. Renault, A new design of electrochemical and optical biosensors based on biocatalytic growth of Au nanoparticles-example of glucose detection, Electroanalysis 25 (2013) 644-651. https://doi.org/10.1002/elan.201200243

[235] B. Sharma, S. Mandani, T.K. Sarma, Enzymes as bionanoreactors: glucose oxidase for the synthesis of catalytic Au nanoparticles and Au nanoparticle-polyaniline nanocomposites, Journal of Materials Chemistry B 2 (2014) 4072-4079. https://doi.org/10.1039/C4TB00218K

[236] T. Yang, Z. Li, L. Wang, C. Guo, Y. Sun, Synthesis, characterization, and self-assembly of protein lysozyme monolayer-stabilized gold nanoparticles, Langmuir 23 (2007) 10533-10538. https://doi.org/10.1021/la701649z

[237] T.T. Baby, S. Ramaprabhu, SiO2 coated Fe3O4 magnetic nanoparticle dispersed multiwalled carbon nanotubes based amperometric glucose biosensor, Talanta 80 (2010) 2016-2022. https://doi.org/10.1016/j.talanta.2009.11.010

[238] P. Díez, R. Villalonga, M.L. Villalonga, J.M. Pingarrón, Supramolecular immobilization of redox enzymes on cyclodextrin-coated magnetic nanoparticles for biosensing applications, Journal of colloid and interface science 386 (2012) 181-188. https://doi.org/10.1016/j.jcis.2012.07.050

[239] M.-H. Liao, J.-C. Guo, W.-C. Chen, A disposable amperometric ethanol biosensor based on screen-printed carbon electrodes mediated with ferricyanide-magnetic

nanoparticle mixture, Journal of Magnetism and Magnetic Materials 304 (2006) e421-e423. https://doi.org/10.1016/j.jmmm.2006.01.223

[240] S. Pal, E.C. Alocilja, Electrically active polyaniline coated magnetic (EAPM) nanoparticle as novel transducer in biosensor for detection of Bacillus anthracis spores in food samples, Biosensors and Bioelectronics 24 (2009) 1437-1444. https://doi.org/10.1016/j.bios.2008.08.020

[241] L. Stanciu, Y.-H. Won, M. Ganesana, S. Andreescu, Magnetic particle-based hybrid platforms for bioanalytical sensors, Sensors 9 (2009) 2976-2999. https://doi.org/10.3390/s90402976

[242] S. Wang, Y. Tan, D. Zhao, G. Liu, Amperometric tyrosinase biosensor based on Fe3O4 nanoparticles-chitosan nanocomposite, Biosensors and Bioelectronics 23 (2008) 1781-1787. https://doi.org/10.1016/j.bios.2008.02.014

[243] W. Putzbach, N.J. Ronkainen, Immobilization techniques in the fabrication of nanomaterial-based electrochemical biosensors: A review, Sensors 13 (2013) 4811-4840. https://doi.org/10.3390/s130404811

[244] T. Endo, R. Ikeda, Y. Yanagida, T. Hatsuzawa, Stimuli-responsive hydrogel-silver nanoparticles composite for development of localized surface plasmon resonance-based optical biosensor, Analytica chimica acta 611 (2008) 205-211. https://doi.org/10.1016/j.aca.2008.01.078

[245] S. Chawla, R. Rawal, C. Pundir, Preparation of cholesterol oxidase nanoparticles and their application in amperometric determination of cholesterol, Journal of nanoparticle research 15 (2013) 1-9. https://doi.org/10.1007/s11051-013-1934-5

[246] O.R. Miranda, X. Li, L. Garcia-Gonzalez, Z.-J. Zhu, B. Yan, U.H. Bunz, V.M. Rotello, Colorimetric bacteria sensing using a supramolecular enzyme-nanoparticle biosensor, Journal of the American Chemical Society 133 (2011) 9650-9653. https://doi.org/10.1021/ja2021729

[247] B. De Strooper, R. Vassar, T. Golde, The secretases: enzymes with therapeutic potential in Alzheimer disease, Nature Reviews Neurology 6 (2010) 99-107. https://doi.org/10.1038/nrneurol.2009.218

[248] A.A. Vertegel, V. Reukov, V. Maximov, Enzyme-Nanoparticle conjugates for biomedical applications, Enzyme Stabilization and Immobilization, Springer2011, pp. 165-182. https://doi.org/10.1007/978-1-60761-895-9_14

[249] M. Wang, J. Zhang, Z. Yuan, W. Yang, Q. Wu, H. Gu, Targeted thrombolysis by using of magnetic mesoporous silica nanoparticles, Journal of biomedical nanotechnology 8 (2012) 624-632. https://doi.org/10.1166/jbn.2012.1416

[250] C.F. Driscoll, R.M. Morris, A.E. Senyei, K.J. Widder, G.S. Heller, Magnetic targeting of microspheres in blood flow, Microvascular research 27 (1984) 353-369. https://doi.org/10.1016/0026-2862(84)90065-7

[251] V. Torchilin, M. Papisov, V. Smirnov, Magnetic Sephadex as a carrier for enzyme immobilization and drug targeting, Journal of biomedical materials research 19 (1985) 461-466. https://doi.org/10.1002/jbm.820190410

[252] C. Capitanescu, A.M. Macovei Oprescu, D. Ionita, G.V. Dinca, C. Turculet, G. Manole, R.A. Macovei, Molecular processes in the streptokinase thrombolytic therapy, Journal of enzyme inhibition and medicinal chemistry 31 (2016) 1411-1414. https://doi.org/10.3109/14756366.2016.1142985

[253] Y.-H. Ma, Y.-W. Hsu, Y.-J. Chang, M.-Y. Hua, J.-P. Chen, T. Wu, Intra-arterial application of magnetic nanoparticles for targeted thrombolytic therapy: A rat embolic model, Journal of magnetism and magnetic materials 311 (2007) 342-346. https://doi.org/10.1016/j.jmmm.2006.10.1204

[254] S.D. Tiukinhoy-Laing, S. Huang, M. Klegerman, C.K. Holland, D.D. McPherson, Ultrasound-facilitated thrombolysis using tissue-plasminogen activator-loaded echogenic liposomes, Thrombosis research 119 (2007) 777-784. https://doi.org/10.1016/j.thromres.2006.06.009

[255] P. Sharma, A.B. Jha, R.S. Dubey, M. Pessarakli, Reactive oxygen species, oxidative damage, and antioxidative defense mechanism in plants under stressful conditions, Journal of botany 2012 (2012). https://doi.org/10.1155/2012/217037

[256] E. Bargagli, C. Olivieri, D. Bennett, A. Prasse, J. Muller-Quernheim, P. Rottoli, Oxidative stress in the pathogenesis of diffuse lung diseases: a review, Respiratory medicine 103 (2009) 1245-1256. https://doi.org/10.1016/j.rmed.2009.04.014

[257] C.R. Kliment, T.D. Oury, Oxidative stress, extracellular matrix targets, and idiopathic pulmonary fibrosis, Free Radical Biology and Medicine 49 (2010) 707-717. https://doi.org/10.1016/j.freeradbiomed.2010.04.036

[258] T.D. Dziubla, V.V. Shuvaev, N.K. Hong, B.J. Hawkins, M. Madesh, H. Takano, E. Simone, M.T. Nakada, A. Fisher, S.M. Albelda, Endothelial targeting of semi-permeable polymer nanocarriers for enzyme therapies, Biomaterials 29 (2008) 215-227. https://doi.org/10.1016/j.biomaterials.2007.09.023

[259] S. Lee, S.C. Yang, M.J. Heffernan, W.R. Taylor, N. Murthy, Polyketal microparticles: a new delivery vehicle for superoxide dismutase, Bioconjugate chemistry 18 (2007) 4-7. https://doi.org/10.1021/bc060259s

[260] M.I. Alam, S. Beg, A. Samad, S. Baboota, K. Kohli, J. Ali, A. Ahuja, M. Akbar, Strategy for effective brain drug delivery, European journal of pharmaceutical sciences 40 (2010) 385-403. https://doi.org/10.1016/j.ejps.2010.05.003

[261] D.J. Begley, Delivery of therapeutic agents to the central nervous system: the problems and the possibilities, Pharmacology & therapeutics 104 (2004) 29-45. https://doi.org/10.1016/j.pharmthera.2004.08.001

[262] V. Reukov, V. Maximov, A. Vertegel, Proteins conjugated to poly (butyl cyanoacrylate) nanoparticles as potential neuroprotective agents, Biotechnology and bioengineering 108 (2011) 243-252. https://doi.org/10.1002/bit.22958

[263] U. Schroeder, P. Sommerfeld, S. Ulrich, B.A. Sabel, Nanoparticle technology for delivery of drugs across the blood-brain barrier, Journal of pharmaceutical sciences 87 (1998) 1305-1307. https://doi.org/10.1021/js980084y

[264] Y.-P. Chen, C.-T. Chen, Y. Hung, C.-M. Chou, T.-P. Liu, M.-R. Liang, C.-T. Chen, C.-Y. Mou, A new strategy for intracellular delivery of enzyme using mesoporous silica nanoparticles: superoxide dismutase, Journal of the American Chemical Society 135 (2013) 1516-1523. https://doi.org/10.1021/ja3105208

[265] H. Yang, L. Qu, A. Wimbrow, X. Jiang, Y.-P. Sun, Enhancing antimicrobial activity of lysozyme against Listeria monocytogenes using immunonanoparticles, Journal of food protection 70 (2007) 1844-1849. https://doi.org/10.4315/0362-028X-70.8.1844

[266] S. Ashraf, M.A. Chatha, W. Ejaz, H.A. Janjua, I. Hussain, Lysozyme-coated silver nanoparticles for differentiating bacterial strains on the basis of antibacterial activity, Nanoscale research letters 9 (2014) 1-10. https://doi.org/10.1186/1556-276X-9-565

[267] R. Satishkumar, A. Vertegel, Charge-directed targeting of antimicrobial protein-nanoparticle conjugates, Biotechnology and bioengineering 100 (2008) 403-412. https://doi.org/10.1002/bit.21782

[268] B. Sharma, A.K. Dangi, P. Shukla, Contemporary enzyme based technologies for bioremediation: a review, Journal of environmental management 210 (2018) 10-22. https://doi.org/10.1016/j.jenvman.2017.12.075

[269] L.F. Stadlmair, T. Letzel, J.E. Drewes, J. Grassmann, Enzymes in removal of pharmaceuticals from wastewater: A critical review of challenges, applications and

screening methods for their selection, Chemosphere 205 (2018) 649-661.
https://doi.org/10.1016/j.chemosphere.2018.04.142

[270] M. Wang, Y. Chen, V.A. Kickhoefer, L.H. Rome, P. Allard, S. Mahendra, A vault-encapsulated enzyme approach for efficient degradation and detoxification of Bisphenol A and its analogues, ACS Sustainable Chemistry & Engineering 7 (2019) 5808-5817. https://doi.org/10.1021/acssuschemeng.8b05432

[271] C.A. Gasser, E.M. Ammann, P. Shahgaldian, P.F.-X. Corvini, Laccases to take on the challenge of emerging organic contaminants in wastewater, Applied microbiology and biotechnology 98 (2014) 9931-9952. https://doi.org/10.1007/s00253-014-6177-6

[272] M. Bilal, M. Adeel, T. Rasheed, Y. Zhao, H.M. Iqbal, Emerging contaminants of high concern and their enzyme-assisted biodegradation-a review, Environment international 124 (2019) 336-353. https://doi.org/10.1016/j.envint.2019.01.011

[273] L. Betancor, H.R. Luckarift, Bioinspired enzyme encapsulation for biocatalysis, Trends in biotechnology 26 (2008) 566-572. https://doi.org/10.1016/j.tibtech.2008.06.009

[274] C.A. Gasser, L. Yu, J. Svojitka, T. Wintgens, E.M. Ammann, P. Shahgaldian, P.F.-X. Corvini, G. Hommes, Advanced enzymatic elimination of phenolic contaminants in wastewater: a nano approach at field scale, Applied microbiology and biotechnology 98 (2014) 3305-3316. https://doi.org/10.1007/s00253-013-5414-8

[275] H.R. Luckarift, M.B. Dickerson, K.H. Sandhage, J.C. Spain, Rapid, room-temperature synthesis of antibacterial bionanocomposites of lysozyme with amorphous silica or titania, Small 2 (2006) 640-643. https://doi.org/10.1002/smll.200500376

[276] P. Asuri, S.S. Karajanagi, R.S. Kane, J.S. Dordick, Polymer-nanotube-enzyme composites as active antifouling films, Small 3 (2007) 50-53. https://doi.org/10.1002/smll.200600312

[277] D. Pan, Y. Gu, H. Lan, Y. Sun, H. Gao, Functional graphene-gold nano-composite fabricated electrochemical biosensor for direct and rapid detection of bisphenol A, Analytica chimica acta 853 (2015) 297-302. https://doi.org/10.1016/j.aca.2014.11.004

[278] R. Pang, M. Li, C. Zhang, Degradation of phenolic compounds by laccase immobilized on carbon nanomaterials: diffusional limitation investigation, Talanta 131 (2015) 38-45. https://doi.org/10.1016/j.talanta.2014.07.045

[279] R.C. Pangule, S.J. Brooks, C.Z. Dinu, S.S. Bale, S.L. Salmon, G. Zhu, D.W. Metzger, R.S. Kane, J.S. Dordick, Antistaphylococcal nanocomposite films based on

enzyme– nanotube conjugates, ACS nano 4 (2010) 3993-4000.
https://doi.org/10.1021/nn100932t

[280] W. Wei, J. Du, J. Li, M. Yan, Q. Zhu, X. Jin, X. Zhu, Z. Hu, Y. Tang, Y. Lu,
Construction of robust enzyme nanocapsules for effective organophosphate
decontamination, detoxification, and protection, Advanced materials 25 (2013) 2212-
2218. https://doi.org/10.1002/adma.201205138

[281] J. Ge, J. Lei, R.N. Zare, Protein-inorganic hybrid nanoflowers, Nature
nanotechnology 7 (2012) 428-432. https://doi.org/10.1038/nnano.2012.80

[282] F. Lyu, Y. Zhang, R.N. Zare, J. Ge, Z. Liu, One-pot synthesis of protein-embedded
metal-organic frameworks with enhanced biological activities, Nano letters 14 (2014)
5761-5765. https://doi.org/10.1021/nl5026419

[283] I. Ocsoy, E. Dogru, S. Usta, A new generation of flowerlike horseradish peroxides
as a nanobiocatalyst for superior enzymatic activity, Enzyme and microbial technology
75 (2015) 25-29. https://doi.org/10.1016/j.enzmictec.2015.04.010

[284] E.P. Cipolatti, A. Valerio, R.O. Henriques, D.E. Moritz, J.L. Ninow, D.M. Freire,
E.A. Manoel, R. Fernandez-Lafuente, D. de Oliveira, Nanomaterials for biocatalyst
immobilization-state of the art and future trends, RSC advances 6 (2016) 104675-
104692. https://doi.org/10.1039/C6RA22047A

[285] J. Liu, Y. Lu, A colorimetric lead biosensor using DNAzyme-directed assembly of
gold nanoparticles, Journal of the American Chemical Society 125 (2003) 6642-6643.
https://doi.org/10.1021/ja034775u

[286] A. Simonian, T. Good, S.-S. Wang, J. Wild, Nanoparticle-based optical biosensors
for the direct detection of organophosphate chemical warfare agents and pesticides,
Analytica chimica acta 534 (2005) 69-77. https://doi.org/10.1016/j.aca.2004.06.056

[287] Y. Azuma, D.L. Bader, D. Hilvert, Substrate sorting by a supercharged
nanoreactor, Journal of the American Chemical Society 140 (2018) 860-863.
https://doi.org/10.1021/jacs.7b11210

[288] D.C. Buehler, M.D. Marsden, S. Shen, D.B. Toso, X. Wu, J.A. Loo, Z.H. Zhou,
V.A. Kickhoefer, P.A. Wender, J.A. Zack, Bioengineered vaults: Self-assembling
protein shell-lipophilic core nanoparticles for drug delivery, ACS nano 8 (2014) 7723-
7732. https://doi.org/10.1021/nn5002694

[289] D.P. Patterson, K. McCoy, C. Fijen, T. Douglas, Constructing catalytic
antimicrobial nanoparticles by encapsulation of hydrogen peroxide producing enzyme

inside the P22 VLP, Journal of Materials Chemistry B 2 (2014) 5948-5951. https://doi.org/10.1039/C4TB00983E

[290] M. Wang, D. Abad, V.A. Kickhoefer, L.H. Rome, S. Mahendra, Vault nanoparticles packaged with enzymes as an efficient pollutant biodegradation technology, ACS nano 9 (2015) 10931-10940. https://doi.org/10.1021/acsnano.5b04073

[291] J. Fu, M. Liu, Y. Liu, N.W. Woodbury, H. Yan, Interenzyme substrate diffusion for an enzyme cascade organized on spatially addressable DNA nanostructures, Journal of the American Chemical Society 134 (2012) 5516-5519. https://doi.org/10.1021/ja300897h

[292] B. Saccà, C.M. Niemeyer, Functionalization of DNA nanostructures with proteins, Chemical Society Reviews 40 (2011) 5910-5921. https://doi.org/10.1039/c1cs15212b

[293] B. Thallinger, E.N. Prasetyo, G.S. Nyanhongo, G.M. Guebitz, Antimicrobial enzymes: an emerging strategy to fight microbes and microbial biofilms, Biotechnology journal 8 (2013) 97-109. https://doi.org/10.1002/biot.201200313

[294] S.H. Bang, A. Jang, J. Yoon, P. Kim, J.S. Kim, Y.-H. Kim, J. Min, Evaluation of whole lysosomal enzymes directly immobilized on titanium (IV) oxide used in the development of antimicrobial agents, Enzyme and microbial technology 49 (2011) 260-265. https://doi.org/10.1016/j.enzmictec.2011.06.004

[295] M. Wang, S.K. Mohanty, S. Mahendra, Nanomaterial-supported enzymes for water purification and monitoring in point-of-use water supply systems, Accounts of chemical research 52 (2019) 876-885. https://doi.org/10.1021/acs.accounts.8b00613

[296] L. Zhu, L. Gong, Y. Zhang, R. Wang, J. Ge, Z. Liu, R.N. Zare, Rapid detection of phenol using a membrane containing laccase nanoflowers, Chemistry-an Asian Journal 8 (2013) 2358-2360. https://doi.org/10.1002/asia.201300020

[297] J. Wang, Amperometric biosensors for clinical and therapeutic drug monitoring: a review, Journal of pharmaceutical and biomedical analysis 19 (1999) 47-53. https://doi.org/10.1016/S0731-7085(98)00056-9

[298] S.R. Couto, J.L.T. Herrera, Industrial and biotechnological applications of laccases: a review, Biotechnology advances 24 (2006) 500-513. https://doi.org/10.1016/j.biotechadv.2006.04.003

[299] A. Kunamneni, F.J. Plou, A. Ballesteros, M. Alcalde, Laccases and their applications: a patent review, Recent patents on biotechnology 2 (2008) 10-24. https://doi.org/10.2174/187220808783330965

[300] S. Riva, Laccases: blue enzymes for green chemistry, TRENDS in Biotechnology 24 (2006) 219-226. https://doi.org/10.1016/j.tibtech.2006.03.006

[301] F. Xu, Applications of oxidoreductases: recent progress, Industrial Biotechnology 1 (2005) 38-50. https://doi.org/10.1089/ind.2005.1.38

[302] H. Claus, G. Faber, H. König, Redox-mediated decolorization of synthetic dyes by fungal laccases, Applied microbiology and biotechnology 59 (2002) 672-678. https://doi.org/10.1007/s00253-002-1047-z

[303] D. Rochefort, D. Leech, R. Bourbonnais, Electron transfer mediator systems for bleaching of paper pulp, Green Chemistry 6 (2004) 14-24. https://doi.org/10.1039/b311898n

Chapter 2

Production, Properties and Applications of Materials-based Nano-Enzymes

A. Fadeyibi

Department of Food and Agricultural Engineering, Faculty of Engineering and Technology, Kwara State University, Malete

adeshina.fadeyibi@kwasu.edu.ng

Abstract

Fungi and bacteria cause foodborne diseases and affect food security, which remains the main challenge of the global food industry. Nanomaterials-based enzyme (NMB) technologies play an important role in improving food security issues. This is possible since they can act quickly and efficiently on food substrates when used as biosensors to monitor and control the quality and shelf life of food. This chapter deals primarily with the applications of NMB in the food industry. The production, properties, and applications of nano-enzymes of carbon, zinc oxide, magnetite, copper, and some noble metals in the food industry were discussed. It was suggested that the material could mimic catalytic activities and compete with other naturally occurring enzymes, such as hydrolase and oxidoreductase in foods. It is hoped that this chapter will provide key insights into NMB technologies applied in the food industry.

Keywords

Fungi and Bacteria, Food Safety, Nano-Enzymes Technology, Biosensor, Food Industry

Contents

1.   Introduction

The NMB technologies are commonly used as biosensors to monitor and control the physical, biological, and chemical activities of food products [1,2]. They may be considered synthetic materials from metals and non-metals, reduced at the nanoscale (1 - 100 nm) [3,4]. Due to their ability to act as a catalyst to accelerate biological and chemical reactions in food products, they are typically used as biosensors for food application [5]. In the past decades, studies have shown that the catalytic behavior of natural enzymes, such as peroxidase and catalase, are mimicked by many biological materials [6,7]. Many novel artificial enzymes have been developed [6,8–11], and when compared with natural enzymes they are more stable, efficient, sensitive and can be easily modified and recycled [6,12]. The materials can also be stored for longer period thereby enhancing their availability for numerous applications [13]. Typically, the NMB are applied in the food industry for studying and controlling the progress of food products deterioration at molecular scale [12,14–18]. The ability of the material to withstand diverse conditions of pressure, pH and temperature makes them suitable for application in monitoring food quality degradation [13,18].

The catalytic behavior of the material, which is responsible for its unique biosensing properties, can be enhanced by improving its surface area or varying the particles size or composition [19]. Food pathogens like fungi and bacteria can be detected, through application of the NMB, using catalytic colorimetric reaction [7,20,21]. The ability of the material to detect the early stage of food product defects is especially the reason they are important in monitoring the yield and market value of crop [22]. However, the design of biosensors made them to act rapidly, reliably, and sensitive to environmental conditions [23]. Although, natural enzymes such as catalase and peroxidase can detect food-borne pathogens, the slow reaction and cost has limited their use [14,23]. There are other proven studies reported on the applications of cost-effective NMB in food processing [4,22,24–26].

The current chapter present the application of the NMB, either in combination with other materials or singly, in the food industry. The in-depth knowledge of production and properties of the material, in relation to their actions against bacteria and fungi in food processing were presented. The applications of nano enzymes of carbon, magnetite, zinc, copper, and other noble metals on food products were discussed, and challenges and prospects were also highlighted.

2.   Production and properties of nanomaterial-based enzymes

The production or synthesis of the NMB are generally carried out by reducing the particle to nano scale size. This often leads to the production of materials that have some unique properties, such as high surface area, low density, and high solubility of solvents at the nanoscale. Significant progress has been made to design new or improve techniques and approaches for the synthesis of the NMB to meet specific industrial applications. The approach used in the production process may be directly related to how they behave in food applications [18]. On this basis, their production can be categorized as physical, chemical, or biological approaches, as described in the Fig. 1 [27–30].

2.1   Chemical synthesis of nanomaterial-based enzymes

The chemical synthesis of the NMB involves the reduction of ions followed by nucleation, by the reaction of the metal with a reducing agent, in the lattice site of the metal [5,31]. The chemical synthesis of the NMB and structures has been described by Shivashankar [32] in his work on nanomaterial synthesis for different applications. The author uses the chemical vapor deposition method and equilibrium thermodynamics to arrive at a nanocomposite coating with interesting features for industrial applications. The chemical method of synthesis of NMB with excellent active properties was reported by Gaspera [33] in her work on wet chemical synthesis of functional nanomaterials. High-quality synthetic enzymes can also be synthesized using a two-phase approach, which entails a significant reduction in the use of chemicals [34]. The enzymes obtained by this approach usually favor catalytic chemical reaction [35], which is essentially needed in the manufacture of biosensors in the food industry.

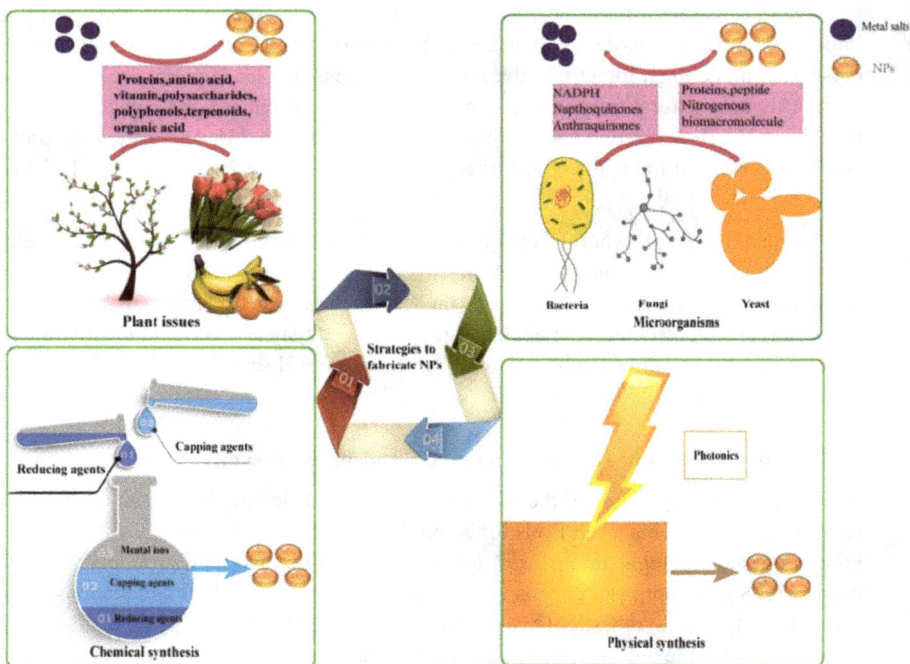

Figure 1: Methods of synthesis of nanomaterials-based enzymes [98]

2.2   Physical synthesis of nanomaterial-based enzymes

The physical synthesis of NMB does not involve the use of chemicals or any biological agent. Several approaches have been reported for enzyme syntheses, laser ablation and condensation methods are common among biosensor manufacturers [19,36]. In the condensation method, the substrate of nanomaterials is normally activated when exposed to temperatures that could render it ionizing and form complex ions that behave like natural enzymes. The method primarily requires adequate space for the combustion activity and a longer duration to complete the pyrolysis or to reach a stable operating temperature [37]. The laser ablation method involves the use of pulse laser irradiation to focus a solid-liquid interface and generate a plasma in the form atoms and ions, which releases quantum of energy in the process [29,38]. This method was used to prepare the high-quality pure NMB [38]. The materials are often associated with a wider distribution of size characteristics that can slow down its catalytic response capacity, thereby retarding its biosensing application. The problem can be resolved by adding a stabilizing agent in the solution to reduce the size of the material and thus improve the surface [38,39]. This approach was applied by Siegel et al. [40] for synthesizing the silver-gold NMB by stabilizing the resulting solution with glycerol.

2.3   Biological synthesis of nanomaterial-based enzymes

Biological synthesis of NMB usually involves the use of living organisms to induce the conversion of materials into synthetic enzymes. Commonly used living organisms include bacteria [41,42], plant [21,39,43] and fungi [44,45], which can act on the main material substrates and reducing it to stable ions [27]. The ions formed are then used to act and destroy the microorganisms that cause spoilage of the food product [27]. In particular, the addition of bacteria into silver nitrate substrates can reduce the silver ion in the cell to nano-silver, which can act as an artificial enzyme in food processing [41,46]. Since fungi can survive extreme conditions of the environment, the NMB synthesized using the fungi has better reactive properties and is more stable than those prepared using the bacteria [47].

Green synthesis is another biological way of producing enzymes based on nanomaterials. [48]. The resulting enzyme will normally have a moderate level of toxicity compared to other biological synthetic methods, as they can incorporate the metabolite into the biological extract to make them biocompatible [48,49]. The plants such as Curcuma longa [50], Euphorbia hirta [51] and Cassia auriculata [21] can convert metallic materials into enzymes, as they can be used as material agents to reduce the main substrate and later stabilize it [27,52,53]. In particular, the Lagenaria siceraria and Malus domestica plant extracts can be used to stabilize and reduce noble metals, like silver and gold, to form NMB suitable for application in the food processing [19,46,54].

2.4   Properties of nanomaterial-based enzymes

The NMB used to detect defects in food quality analysis is characterized by its high reactive surface, which can cause it to disperse [55] and interfere with food structure [56]. These properties enable the materials to enhance catalytic chemical reaction between itself and the food substrates during processing, thereby revealing the areas of defects in the substrates. The enzymes can adsorb and interact with the chemical or biomolecules due to high adsorption [57]. This therefore presents the unique ability of the materials for use as detection sensors [58]. The examples of sensors utilizing the NMB includes but are not limited to colorimetric, electrochemical, and fluorescent sensors. Other classes of synthetic enzymes that are useful for detection of estradiol in red wine include silver and gold NMB

[59]. This depends to a large extent on the size of the nanoparticles, which can be improved further to meet other industrial applications [58].

3. Application of nanomaterial-based enzymes in the food industry

3.1 Carbon-based nanomaterial enzyme biosensors

The ability of the carbon-based NMB biosensors to control the activities of deteriorating microorganisms in food and water, thereby enhancing the quality of food and water in the supply chain, have been studied [60,61]. For example, Rao et al. [62] reviewed the numerous potentials of carbon nanotubes, buckminsterfullerene, activated carbon, and graphene oxide for application in food preservation, water purification and other applications. Also, Abdelmonem [63] reported the use of the carbon-based NMB to monitor and control how thermoplastic films can prevent food spoilage and extend its shelf-life. The materials are also used to detect contaminants in liquid foods and to identify or separate spoiled products from many industrial foods [64]. The nanomaterials can in fact be used to control the activities of certain biological reactions that may affect yield in food production [56,64,65]. The incidence of tobacco mosaic virus infection can be suppressed using the nanocarbon biosensor [66]. The protective role of the nanomaterial was made possible by the physical mobility of the carbon ions that attack the virus cells to prevent its further replication [66], as shown in Fig. 2. The aquatic ecotoxicity of wastewater has also been treated using the nanocarbon based enzymes with a great degree of success [67].

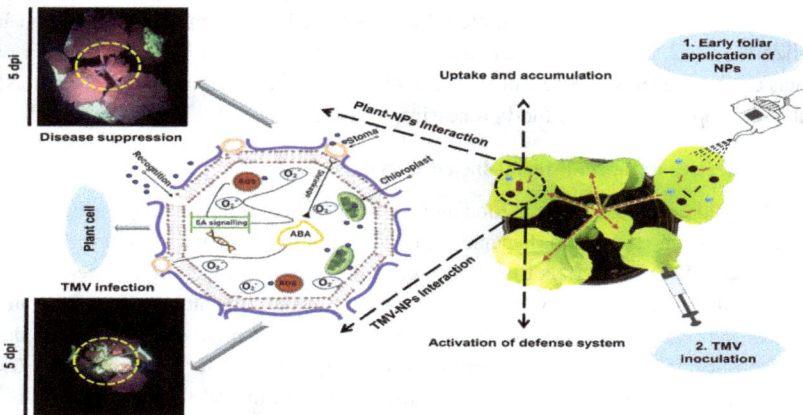

Figure 2: The use of nano carbon-based enzyme to suppress tobacco mosaic virus [66]

3.2   Zinc oxide-based nanomaterial enzyme biosensors

The application of zinc oxide NMB as biosensors to monitor and control postharvest operations in the food sector has been extensively reported in literature [16,68–75]. Akbar and Sadiq [76] applied the zinc-oxide nanomaterials to control the microbial activities in food product. The antimicrobial potential of the zinc-oxide NMB can be associated with the ability of the materials to hydrolyze water to form a collection of hydroxyl ions [77]. This, in turn, will act on the food substrates to target the activities of the fungi and bacterial causing damage to food [25]. Also, the material, when incorporated with acetylcholinesterase, forms an active biosensing product for the detection of pesticides in storage facilities, such as silos, and bans [78]. This biosensor can detect paraoxon pesticide in the range 0.035–1.38 ppm and can be used to detect other acetylcholinesterase inhibiting organophosphate pesticides in stored food products. The material has also been used for detecting and monitoring the extend of adulteration in milk and milk product in the food industry [17]. The detection of analyte in milk using the nano-zinc enzyme biosensor can be visualized in Fig. 3 [17].

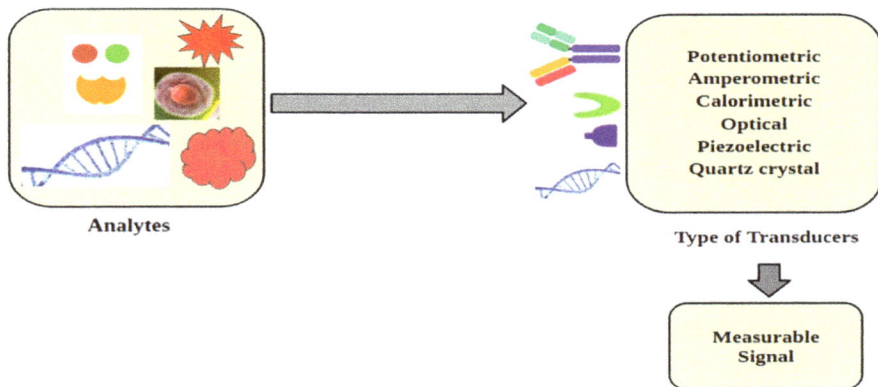

Figure 3: Nano-zinc based enzyme biosensor for detecting analytes in milk [17]

3.3   Magnetite-based nanomaterial enzyme biosensors

The mechanism of magnetite NMB has been explained by Wang and Gunasekaran[15] in their study on artificial enzymes for safety and quality of food. The authors viewed the magnetite nanoparticle as a close replica of the peroxidase natural enzymes with similar properties. We can see how, in Fig 4, the food substrates such as the 3,30,5,50-tetramethylbenzidine (TMB), o-phenylenediamine (OPD), 2,2-azinobis and 10-acetyl-3

(ABTS), and 7-dihydroxyphenoxazine (AR) are oxidized in the presence of the magnetite-based nanomaterials enzymes [15]. This closely imitates the behavior of the oxidase and peroxidase natural enzymes. Other investigators such as Nie et al.[79], Zhang et al. [6] and Gao et al. [8] confirmed the close behaviour of the magnetite NMB with the peroxidase in biosensing applications. Also, the findings of Wan et al. [10] and Wang et al. [11] on the catalytic experiments and steady-state kinetic of the material suggest that the particles undergo enzymatic catalysis just like the natural enzyme peroxidase. The catalytic mechanism was explained by Zhang et al. [9] and Wang [80] to involve the reaction of a collection of hydroxyl ions obtained from the exposure of Fe^{2+} to the hydrogen peroxide. The hydroxyl ion will immediately bind with the hydrogen ion to cause an oxidation of the Fe^{2+} to Fe^{3+}, and the circle repeat itself again. This is how the magnetite NMB achieve its peroxidase like properties, and by this, has often been used as surrogate artificial enzyme in the food industry. The application of the material in food analysis has been extensively studied [15,26,64,81,82]. Other applications of the nano-enzyme have been reported for wastewater and effluents treatment from the food industry [73,83–85].

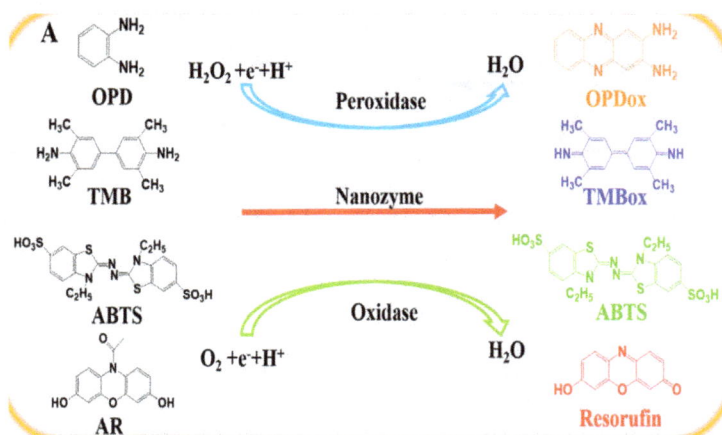

Figure 4: Oxidation reaction of oxidase and peroxidate materials in the presence of magetite nanoparticles and oxidizing agents [15]

3.4   Copper cluster-based nanomaterial enzyme biosensors

The coper nanocluster NMB exhibits impressive catalytic oxidation reaction between hydrogen peroxide and the substrates. This is due to the ability of copper to exist in two atomic forms in its core, namely five and thirteen, thus indicating different oxidation levels [86]. In the same way as magnetic enzyme biosensors, copper nanoclusters exhibit a higher enzymic behavior than catalase and peroxidase [87]. The reactive mechanism here involves the copper nanoclusters forming bonds with the O-O and the hydrogen peroxide groups to form a substrate and catalytic active sites, which are then rapidly hydrolysed to trigger the oxidation of the substrate. The catalytic response mechanism imitating peroxidase and oxidase enzymes in the oxidative action of the nanomaterial on the food substrate can be seen in the Fig. 5 [7].

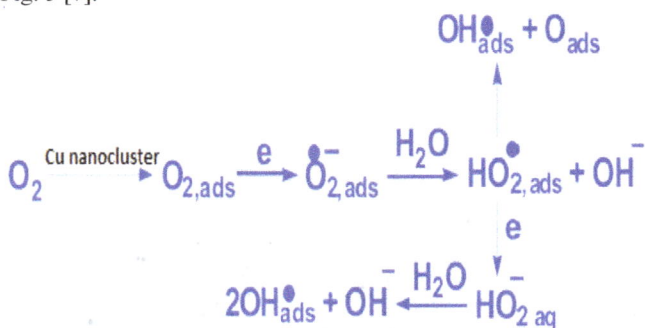

$$OH^{\bullet}_{ads} + O_{ads}$$

$$O_2 \xrightarrow{Cu\ nanocluster} O_{2,ads} \xrightarrow{e} O^{\bullet-}_{2,ads} \xrightarrow{H_2O} HO^{\bullet}_{2,ads} + OH^-$$

$$2OH^{\bullet}_{ads} + OH^- \xleftarrow{H_2O} HO^-_{2\,aq} \xleftarrow{e}$$

Figure 5: Mechanism of catalytic reaction of copper nanocluster [7]

The copper nanocluster can be used to detect hydrogen peroxide and glucose in food analysis. The action of the NMB here resembles that of peroxidase. The material has been applied in the study of glucose degradation and glucose detection analysis in biomedical and food industries [87]. Other applications of the material are found in the treatment of wastewater and in detecting organic contaminants [2]. Some of the physical properties of the copper-based nanomaterial enzyme which favors its application for wastewater and contaminant treatments include its ability to disperse widely in water, biocompactibility and its high stability when exposed to sunlight [88,89]. Structurally, Bhamore et al. [90] noted that the copper nanocluster has a high affinity to identify or recognize ligand molecules due probably to its transition nature, thus they are suitable for detecting contaminants in wastewater. The detection capability of the NMB can be extended to small ions and internal microstructural defects in food and other applications [2,91–93].

3.5 Noble metal-based nanomaterial enzyme biosensors

The noble metal NMB, which include silver, platinum, and lead ion-based enzyme sensors, can act on food substrates to induce catalytic chemical reactions under wide range of experimental conditions [20]. Although, these materials may show promising abilities in biosensing applications (Fig. 6), they often have a tendency for short shelf-life and may be unstable when expose to extreme conditions. For this reason, they are often used together with stabilizers to help prolong the life span and maintain the stability in food preservation. The combine effect of the materials and the stabilizers on the food substrates has been noted to effectively promote catalytic chemical reaction just like the redox reaction of the natural enzyme peroxidase in fruits and vegetables [94]. For example, the gold nanocluster combined with bovine serum albumin was reported to present high peroxidase-like behavior in wide temperature range and offers improved physical properties than ordinary gold nanocluster biosensors in food processing [12,95].

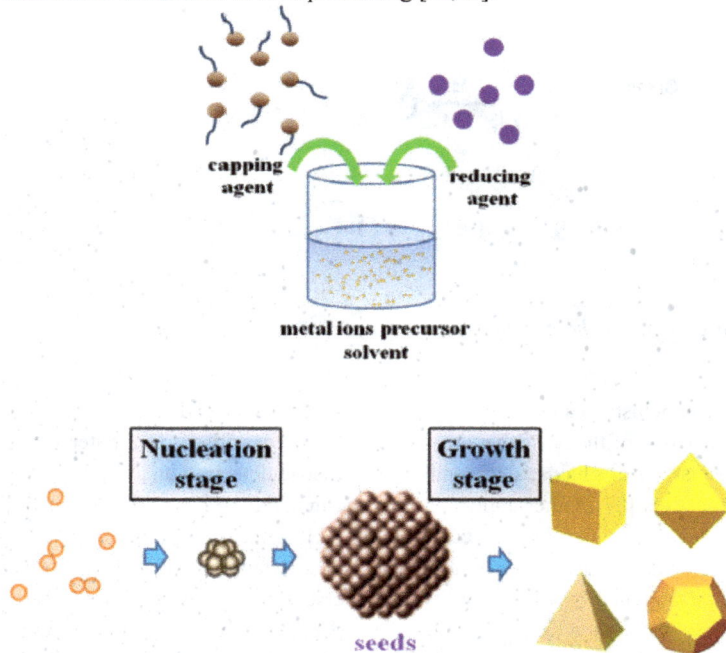

Figure 6: Illustration of synthesis of noble metals nano-enzymes [20]

Materials Research Forum LLC
https://doi.org/10.21741/9781644901977-2

The application of the noble metal NMB for oxygen scavenging in the modified atmosphere packaging materials maybe possible by hydro-deoxygenation reaction [96]. The material can introduce hyroxyl ions in the storage atmosphere that will inturn react with the excess oxygen to create suitable environment for keeping the food [16,72]. The trimetallic catalytic reaction of the noble metals might be responsible for their unique ability as sensing materials for oxygen. This reaction can also be extended to other types of packaging designs that influence the gaseous composition of the storage atmosphere [14,97]. Also, the noble metals nano-enzymes can be used for detecting incidence of food spoilage due to pest, rodents, or insect attacks [98].

4.   Challenges and prospects

After extensive literature review, we noticed that the utilization of NMB in the food industry is currently limited to closely imitating the behavior of natural enzymes, such as oxidase and peroxidase. These two enzymes are the driving forces in the catalytic chemical reactions and changes occurring in food processing and preservation. The natural enzymes are however affected by pH and temperature conditions, which may either speedup or reduce their action on the food substrates. Meanwhile, the nanomaterials are designed to address this deficit of the natural enzymes by providing enzyme like artificial NMBs that can withstand extreme environmental conditions, like pH and temperature. Further research should focus on designing NMB that can mimic the catalytic activities and compete with other natural enzymes including hydrolases, oxidoreductase, transferases, isomerases, and ligases present in foods.

Conclusions

The use of biosensor for the detection of various food pathogens can be traced back to decades, and many of them that are being used nowadays have come from NMB. Because of unique inherent properties, the NMB enzymes have continued to be one of the major bio-sensing technologies still today. In the recent years, there has been a remarkable resurgence of interests observed in the application of the materials in the food industry. With significance usage in monitoring and control of food pathogens, the NMB enzymes have gain new momentum. They are now suitable for effective screening of pathogens in foods. Interaction of the materials with chemical or biomolecules, due to the high adsorption, is one of their main properties, especially in relation to actions against food-borne pathogens. Methods of production and properties of biosensors from various NMB enzymes are available in the literature; the advances and sensitivity of the materials have made their interaction less laborious and easy to apply. The current chapter presents the application of NMB enzymes in the food industry.

References

[1] F. Mustafa, A. Othman, S. Andreescu, Cerium oxide-based hypoxanthine biosensor for Fish spoilage monitoring, Sensors and Actuators, B: Chemical. 332 (2021) 129435. https://doi.org/10.1016/j.snb.2021.129435

[2] Y.-S. Lin, Y.-F. Lin, A. Nain, Y.-F. Huang, H.-T. Chang, A critical review of copper nanoclusters for monitoring of water quality, Sensors and Actuators Reports. 3 (2021) 100026. https://doi.org/10.1016/j.snr.2021.100026

[3] W. Wang, X. Wang, N. Cheng, Y. Luo, Y. Lin, W. Xu, D. Du, Recent advances in nanomaterials-based electrochemical (bio)sensors for pesticides detection, TrAC - Trends in Analytical Chemistry. 132 (2020) 116041. https://doi.org/10.1016/j.trac.2020.116041

[4] W.A. Khan, M.B. Arain, M. Soylak, Nanomaterials-based solid phase extraction and solid phase microextraction for heavy metals food toxicity, Food and Chemical Toxicology. 145 (2020) 111704. https://doi.org/10.1016/j.fct.2020.111704

[5] J.B. Deshpande, A.A. Kulkarni, Reaction Engineering for Continuous Production of Silver Nanoparticles, Chemical Engineering and Technology. 41 (2018) 157-167. https://doi.org/10.1002/ceat.201700035

[6] Z. Zhang, X. Wang, X. Yang, A sensitive choline biosensor using Fe3O4 magnetic nanoparticles as peroxidase mimics, Analyst. 136 (2011) 4960-4965. https://doi.org/10.1039/c1an15602k

[7] S. Cai, C. Qi, Y. Li, Q. Han, R. Yang, C. Wang, PtCo bimetallic nanoparticles with high oxidase-like catalytic activity and their applications for magnetic-enhanced colorimetric biosensing, Journal of Materials Chemistry B. 4 (2016) 1869-1877. https://doi.org/10.1039/C5TB02052B

[8] L. Gao, J. Zhuang, L. Nie, J. Zhang, Y. Zhang, N. Gu, T. Wang, J. Feng, D. Yang, S. Perrett, X. Yan, Intrinsic peroxidase-like activity of ferromagnetic nanoparticles, Nature Nanotechnology. 2 (2007) 577-583. https://doi.org/10.1038/nnano.2007.260

[9] X.Q. Zhang, S.W. Gong, Y. Zhang, T. Yang, C.Y. Wang, N. Gu, Prussian blue modified iron oxide magnetic nanoparticles and their high peroxidase-like activity, Journal of Materials Chemistry. 20 (2010) 5110-5116. https://doi.org/10.1039/c0jm00174k

[10] D. Wan, W. Li, G. Wang, X. Wei, Shape-Controllable Synthesis of Peroxidase-Like Fe3O4 Nanoparticles for Catalytic Removal of Organic Pollutants, Journal of

Materials Engineering and Performance. 25 (2016) 4333-4340.
https://doi.org/10.1007/s11665-016-2283-1

[11] N. Wang, L. Zhu, D. Wang, M. Wang, Z. Lin, H. Tang, Sono-assisted preparation of highly efficient peroxidase-like Fe3O4 magnetic nanoparticles for catalytic removal of organic pollutants with H2O2, Ultrasonics Sonochemistry. 17 (2010) 526-533. https://doi.org/10.1016/j.ultsonch.2009.11.001

[12] X.X. Wang, Q. Wu, Z. Shan, Q.M. Huang, BSA-stabilized Au clusters as peroxidase mimetics for use in xanthine detection, Biosensors and Bioelectronics. 26 (2011) 3614-3619. https://doi.org/10.1016/j.bios.2011.02.014

[13] S. Chaudhary, A. Umar, S.K. Mehta, Selenium nanomaterials: An overview of recent developments in synthesis, properties and potential applications, Progress in Materials Science. 83 (2016) 270-329. https://doi.org/10.1016/j.pmatsci.2016.07.001

[14] G.S. Birth, G. Eisler, Optics For Food Quality Analysis, in: Optics in Quality Assurance II, SPIE, 1971: pp. 17-17.

[15] W. Wang, S. Gunasekaran, Nanozymes-based biosensors for food quality and safety, TrAC - Trends in Analytical Chemistry. 126 (2020) 115841. https://doi.org/10.1016/j.trac.2020.115841

[16] A. Fadeyibi, Z. Osunde, M. Yisa, Optimization of processing parameters of nanocomposite film for fresh sliced okra packaging, Journal of Applied Packaging Research. 11 (2019). https://scholarworks.rit.edu/japr/vol11/iss2/1 (accessed June 29, 2021).

[17] R. Nagraik, A. Sharma, D. Kumar, P. Chawla, A.P. Kumar, Milk adulterant detection: Conventional and biosensor-based approaches: A review, Sensing and Bio-Sensing Research. 33 (2021) 100433. https://doi.org/10.1016/j.sbsr.2021.100433

[18] J. Heck, J. Goding, R. Portillo Lara, R. Green, The influence of physicochemical properties on the processibility of conducting polymers: A bioelectronics perspective, Acta Biomaterialia. (2021). https://doi.org/10.1016/j.actbio.2021.05.052

[19] L. Wei, J. Lu, H. Xu, A. Patel, Z.S. Chen, G. Chen, Silver nanoparticles: Synthesis, properties, and therapeutic applications, Drug Discovery Today. 20 (2015) 595-601. https://doi.org/10.1016/j.drudis.2014.11.014

[20] W. Zang, G. Li, L. Wang, X. Zhang, Catalytic hydrogenation by noble-metal nanocrystals with well-defined facets: A review, Catalysis Science and Technology. 5 (2015) 2532-2553. https://doi.org/10.1039/C4CY01619J

[21] K. Muthu, S. Priya, Green synthesis, characterization and catalytic activity of silver nanoparticles using Cassia auriculata flower extract separated fraction, Spectrochimica Acta - Part A: Molecular and Biomolecular Spectroscopy. 179 (2017) 66-72. https://doi.org/10.1016/j.saa.2017.02.024

[22] N. Chaudhry, S. Dwivedi, V. Chaudhry, A. Singh, Q. Saquib, A. Azam, J. Musarrat, Bio-inspired nanomaterials in agriculture and food: Current status, foreseen applications and challenges, Microbial Pathogenesis. 123 (2018) 196-200. https://doi.org/10.1016/j.micpath.2018.07.013

[23] S.K. Ameta, A.K. Rai, D. Hiran, R. Ameta, S.C. Ameta, Use of nanomaterials in food science, in: Biogenic Nanoparticles and Their Use in Agro-Ecosystems, Springer Singapore, 2020: pp. 457-488. https://doi.org/10.1007/978-981-15-2985-6_24

[24] Z. Mohammadi, S.M. Jafari, Detection of food spoilage and adulteration by novel nanomaterial-based sensors, Advances in Colloid and Interface Science. 286 (2020) 102297. https://doi.org/10.1016/j.cis.2020.102297

[25] R. Gupta, N. Raza, S.K. Bhardwaj, K. Vikrant, K.H. Kim, N. Bhardwaj, Advances in nanomaterial-based electrochemical biosensors for the detection of microbial toxins, pathogenic bacteria in food matrices, Journal of Hazardous Materials. 401 (2021) 123379. https://doi.org/10.1016/j.jhazmat.2020.123379

[26] M. Faraji, Y. Yamini, Application of magnetic nanomaterials in food analysis, in: Magnetic Nanomaterials in Analytical Chemistry, Elsevier, 2021: pp. 87-120. https://doi.org/10.1016/B978-0-12-822131-0.00003-0

[27] P. Singh, Y.J. Kim, D. Zhang, D.C. Yang, Biological Synthesis of Nanoparticles from Plants and Microorganisms, Trends in Biotechnology. 34 (2016) 588-599. https://doi.org/10.1016/j.tibtech.2016.02.006

[28] M.A. Faramarzi, A. Sadighi, Insights into biogenic and chemical production of inorganic nanomaterials and nanostructures, Advances in Colloid and Interface Science. 189-190 (2013) 1-20. https://doi.org/10.1016/j.cis.2012.12.001

[29] D. Zhang, B. Gökce, S. Barcikowski, Laser Synthesis and Processing of Colloids: Fundamentals and Applications, Chemical Reviews. 117 (2017) 3990-4103. https://doi.org/10.1021/acs.chemrev.6b00468

[30] M.T. Alula, L. Karamchand, N.R. Hendricks, J.M. Blackburn, Citrate-capped silver nanoparticles as a probe for sensitive and selective colorimetric and spectrophotometric sensing of creatinine in human urine, Analytica Chimica Acta. 1007 (2018) 40-49. https://doi.org/10.1016/j.aca.2017.12.016

[31] T. Oku, T. Kusunose, K. Niihara, K. Suganuma, Chemical synthesis of silver nanoparticles encapsulated in boron nitride nanocages, Journal of Materials Chemistry. 10 (2000) 255-257. https://doi.org/10.1039/a908351k

[32] S.A. Shivashankar, Chemical synthesis of nanomaterials and structures, including nanostructured thin films, for different applications, Springer Tracts in Mechanical Engineering. 14 (2014) 249-263. https://doi.org/10.1007/978-81-322-1913-2_15

[33] E. della Gaspera, Special issue "wet chemical synthesis of functional nanomaterials," Nanomaterials. 11 (2021). https://doi.org/10.3390/nano11041044

[34] H. Duan, D. Wang, Y. Li, Green chemistry for nanoparticle synthesis, Chemical Society Reviews. 44 (2015) 5778-5792. https://doi.org/10.1039/C4CS00363B

[35] M. Khalil, G.T.M. Kadja, M.M. Ilmi, Advanced nanomaterials for catalysis: Current progress in fine chemical synthesis, hydrocarbon processing, and renewable energy, Journal of Industrial and Engineering Chemistry. 93 (2021) 78-100. https://doi.org/10.1016/j.jiec.2020.09.028

[36] J.H. Jung, H. Cheol Oh, H. Soo Noh, J.H. Ji, S. Soo Kim, Metal nanoparticle generation using a small ceramic heater with a local heating area, Journal of Aerosol Science. 37 (2006) 1662-1670. https://doi.org/10.1016/j.jaerosci.2006.09.002

[37] M.H. Magnusson, K. Deppert, J.O. Malm, J.O. Bovin, L. Samuelson, Gold nanoparticles: Production, reshaping, and thermal charging, Journal of Nanoparticle Research. 1 (1999) 243-251. https://doi.org/10.1023/A:1010012802415

[38] M. Sakamoto, M. Fujistuka, T. Majima, Light as a construction tool of metal nanoparticles: Synthesis and mechanism, Journal of Photochemistry and Photobiology C: Photochemistry Reviews. 10 (2009) 33-56. https://doi.org/10.1016/j.jphotochemrev.2008.11.002

[39] I.M. El-Sherbiny, A. El-Shibiny, E. Salih, Photo-induced green synthesis and antimicrobial efficacy of poly (ϵ-caprolactone)/curcumin/grape leaf extract-silver hybrid nanoparticles, Journal of Photochemistry and Photobiology B: Biology. 160 (2016) 355-363. https://doi.org/10.1016/j.jphotobiol.2016.04.029

[40] J. Siegel, O. Kvítek, P. Ulbrich, Z. Kolská, P. Slepička, V. Švorčík, Progressive approach for metal nanoparticle synthesis, Materials Letters. 89 (2012) 47-50. https://doi.org/10.1016/j.matlet.2012.08.048

[41] M. Saravanan, S. Arokiyaraj, T. Lakshmi, A. Pugazhendhi, Synthesis of silver nanoparticles from Phenerochaete chrysosporium (MTCC-787) and their antibacterial

activity against human pathogenic bacteria, Microbial Pathogenesis. 117 (2018) 68-72. https://doi.org/10.1016/j.micpath.2018.02.008

[42] Z. Chen, Z. Li, G. Chen, J. Zhu, Q. Liu, T. Feng, In situ formation of AgNPs on S. cerevisiae surface as bionanocomposites for bacteria killing and heavy metal removal, International Journal of Environmental Science and Technology. 14 (2017) 1635-1642. https://doi.org/10.1007/s13762-017-1261-y

[43] V. Kumar, D. Bano, S. Mohan, D.K. Singh, S.H. Hasan, Sunlight-induced green synthesis of silver nanoparticles using aqueous leaf extract of Polyalthia longifolia and its antioxidant activity, Materials Letters. 181 (2016) 371-377. https://doi.org/10.1016/j.matlet.2016.05.097

[44] S.M. El-Sonbaty, Fungus-mediated synthesis of silver nanoparticles and evaluation of antitumor activity, Cancer Nanotechnology. 4 (2013) 73-79. https://doi.org/10.1007/s12645-013-0038-3

[45] L. Ma, W. Su, J.X. Liu, X.X. Zeng, Z. Huang, W. Li, Z.C. Liu, J.X. Tang, Optimization for extracellular biosynthesis of silver nanoparticles by Penicillium aculeatum Su1 and their antimicrobial activity and cytotoxic effect compared with silver ions, Materials Science and Engineering C. 77 (2017) 963-971. https://doi.org/10.1016/j.msec.2017.03.294

[46] X. Zhao, L. Zhou, M.S. Riaz Rajoka, L. Yan, C. Jiang, D. Shao, J. Zhu, J. Shi, Q. Huang, H. Yang, M. Jin, Fungal silver nanoparticles: synthesis, application and challenges, Critical Reviews in Biotechnology. 38 (2018) 817-835. https://doi.org/10.1080/07388551.2017.1414141

[47] M. Kitching, P. Choudhary, S. Inguva, Y. Guo, M. Ramani, S.K. Das, E. Marsili, Fungal surface protein mediated one-pot synthesis of stable and hemocompatible gold nanoparticles, Enzyme and Microbial Technology. 95 (2016) 76-84. https://doi.org/10.1016/j.enzmictec.2016.08.007

[48] P. Aarthye, M. Sureshkumar, Green synthesis of nanomaterials: An overview, Materials Today: Proceedings. (2021). https://doi.org/10.1016/j.matpr.2021.04.564

[49] B. Amanulla, H.K.R. Subbu, S.K. Ramaraj, A sonochemical synthesis of cyclodextrin functionalized Au-FeNPs for colorimetric detection of Cr6+ in different industrial wastewater, Ultrasonics Sonochemistry. 42 (2018) 747-753. https://doi.org/10.1016/j.ultsonch.2017.12.041

[50] R. Sankar, P.K.S.M. Rahman, K. Varunkumar, C. Anusha, A. Kalaiarasi, K.S. Shivashangari, V. Ravikumar, Facile synthesis of Curcuma longa tuber powder

engineered metal nanoparticles for bioimaging applications, Journal of Molecular Structure. 1129 (2017) 8-16. https://doi.org/10.1016/j.molstruc.2016.09.054

[51] V. Kumar, D.K. Singh, S. Mohan, R.K. Gundampati, S.H. Hasan, Photoinduced green synthesis of silver nanoparticles using aqueous extract of Physalis angulata and its antibacterial and antioxidant activity, Journal of Environmental Chemical Engineering. 5 (2017) 744-756. https://doi.org/10.1016/j.jece.2016.12.055

[52] A. Mohammed Fayaz, M. Girilal, R. Venkatesan, P.T. Kalaichelvan, Biosynthesis of anisotropic gold nanoparticles using Maduca longifolia extract and their potential in infrared absorption, Colloids and Surfaces B: Biointerfaces. 88 (2011) 287-291. https://doi.org/10.1016/j.colsurfb.2011.07.003

[53] C.K. Tagad, S.R. Dugasani, R. Aiyer, S. Park, A. Kulkarni, S. Sabharwal, Green synthesis of silver nanoparticles and their application for the development of optical fiber-based hydrogen peroxide sensor, Sensors and Actuators, B: Chemical. 183 (2013) 144-149. https://doi.org/10.1016/j.snb.2013.03.106

[54] A.M. Baetsen-Young, M. Vasher, L.L. Matta, P. Colgan, E.C. Alocilja, B. Day, Direct colorimetric detection of unamplified pathogen DNA by dextrin-capped gold nanoparticles, Biosensors and Bioelectronics. 101 (2018) 29-36. https://doi.org/10.1016/j.bios.2017.10.011

[55] A.K. Srivastava, A. Dev, S. Karmakar, Nanosensors and nanobiosensors in food and agriculture, Environmental Chemistry Letters. 16 (2018) 161-182. https://doi.org/10.1007/s10311-017-0674-7

[56] H. Ahangari, S. Kurbanoglu, A. Ehsani, B. Uslu, Latest trends for biogenic amines detection in foods: Enzymatic biosensors and nanozymes applications, Trends in Food Science & Technology. 112 (2021) 75-87. https://doi.org/10.1016/j.tifs.2021.03.037

[57] G. Maduraiveeran, M. Sasidharan, V. Ganesan, Electrochemical sensor and biosensor platforms based on advanced nanomaterials for biological and biomedical applications, Biosensors and Bioelectronics. 103 (2018) 113-129. https://doi.org/10.1016/j.bios.2017.12.031

[58] M. Waqas, A. Zulfiqar, H.B. Ahmad, N. Akhtar, M. Hussain, Z. Shafiq, Y. Abbas, K. Mehmood, M. Ajmal, M. Yang, Fabrication of highly stable silver nanoparticles with shape-dependent electrochemical efficacy, Electrochimica Acta. 271 (2018) 641-651. https://doi.org/10.1016/j.electacta.2018.03.049

[59] J.H. Soh, Y. Lin, S. Rana, J.Y. Ying, M.M. Stevens, Colorimetric Detection of Small Molecules in Complex Matrixes via Target-Mediated Growth of Aptamer-

Functionalized Gold Nanoparticles, Analytical Chemistry. 87 (2015) 7644-7652.
https://doi.org/10.1021/acs.analchem.5b00875

[60] M. Azizi-Lalabadi, H. Hashemi, J. Feng, S.M. Jafari, Carbon nanomaterials against
pathogens; the antimicrobial activity of carbon nanotubes, graphene/graphene oxide,
fullerenes, and their nanocomposites, Advances in Colloid and Interface Science. 284
(2020) 102250. https://doi.org/10.1016/j.cis.2020.102250

[61] L. Giraud, A. Tourrette, E. Flahaut, Carbon nanomaterials-based polymer-matrix
nanocomposites for antimicrobial applications: A review, Carbon. 182 (2021) 463-
483. https://doi.org/10.1016/j.carbon.2021.06.002

[62] N. Rao, R. Singh, L. Bashambu, Carbon-based nanomaterials: Synthesis and
prospective applications, in: Materials Today: Proceedings, Elsevier Ltd, 2021: pp.
608-614. https://doi.org/10.1016/j.matpr.2020.10.593

[63] A.M. Abdelmonem, Application of Carbon-Based Nanomaterials in Food
Preservation Area, in: Carbon Nanomaterials for Agri-Food and Environmental
Applications, Elsevier, 2019: pp. 583-593. https://doi.org/10.1016/B978-0-12-819786-
8.00025-6

[64] S. Eyvazi, B. Baradaran, A. Mokhtarzadeh, M. de la Guardia, Recent advances on
development of portable biosensors for monitoring of biological contaminants in
foods, Trends in Food Science & Technology. 114 (2021) 712-721.
https://doi.org/10.1016/j.tifs.2021.06.024

[65] E.S. McLamore, E. Alocilja, C. Gomes, S. Gunasekaran, D. Jenkins, S.P.A. Datta,
Y. Li, Y. (Jessie) Mao, S.R. Nugen, J.I. Reyes-De-Corcuera, P. Takhistov, O. Tsyusko,
J.P. Cochran, T.R. (Jeremy) Tzeng, J.Y. Yoon, C. Yu, A. Zhou, FEAST of biosensors:
Food, environmental and agricultural sensing technologies (FEAST) in North
America, Biosensors and Bioelectronics. 178 (2021) 113011.
https://doi.org/10.1016/j.bios.2021.113011

[66] M. Adeel, T. Farooq, J.C. White, Y. Hao, Z. He, Y. Rui, Carbon-based
nanomaterials suppress tobacco mosaic virus (TMV) infection and induce resistance in
Nicotiana benthamiana, Journal of Hazardous Materials. 404 (2021) 124167.
https://doi.org/10.1016/j.jhazmat.2020.124167

[67] C. Zhang, X. Chen, S.H. Ho, Wastewater treatment nexus: Carbon nanomaterials
towards potential aquatic ecotoxicity, Journal of Hazardous Materials. 417 (2021)
125959. https://doi.org/10.1016/j.jhazmat.2021.125959

Nanomaterial-Supported Enzymes   Materials Research Forum LLC
Materials Research Foundations **126** (2022) 67-88   https://doi.org/10.21741/9781644901977-2

[68] M. Ahmar Rauf, M. Oves, F. Ur Rehman, A. Rauf Khan, N. Husain, Bougainvillea flower extract mediated zinc oxide's nanomaterials for antimicrobial and anticancer activity, Biomedicine and Pharmacotherapy. 116 (2019) 108983. https://doi.org/10.1016/j.biopha.2019.108983

[69] S. Vijayakumar, E. Vidhya, M. Nilavukkarasi, V.N. Punitha, P.K. Praseetha, Potential eco-friendly Zinc Oxide nanomaterials through Phyco-nanotechnology -A review, Biocatalysis and Agricultural Biotechnology. 35 (2021) 102050. https://doi.org/10.1016/j.bcab.2021.102050

[70] S.P.K. Malhotra, Applications of zinc oxide nanoparticles as an antimicrobial agent in the food packaging industry, in: Zinc-Based Nanostructures for Environmental and Agricultural Applications, Elsevier, 2021: pp. 125-137. https://doi.org/10.1016/B978-0-12-822836-4.00021-5

[71] C.M. Hussain, R. Keçili, Use of nanomaterials for environmental analysis, in: Modern Environmental Analysis Techniques for Pollutants, Elsevier, 2020: pp. 277-322. https://doi.org/10.1016/B978-0-12-816934-6.00011-4

[72] A. Fadeyibi, Z. Osunde, M. Yisa, Effects of period and temperature on quality and shelf-life of cucumber and garden-eggs packaged using cassava starch-zinc nanocomposite film, J. Appl. Packag. Res. 12 (2020) 3.

[73] A. Fadeyibi, M.G. Yisa, F.A. Adeniji, K.K. Katibi, K.P. Alabi, K.R. Adebayo, Potentials of zinc and magnetite nanoparticles for contaminated water treatment, Agricultural Reviews. 39 (1) (2018) 175- 180. https://doi.org/10.18805/ag.R-113

[74] A. Vasilescu, C. Polonschii, A.M. Titoiu, R. Mishra, S. Peteu, J.-L. Marty, Bioassays and biosensors for food analysis: focus on allergens and food packaging, Commercial Biosensors and Their Applications. (2020) 217-258. https://doi.org/10.1016/B978-0-12-818592-6.00009-8

[75] C. Griesche, A.J. Baeumner, Biosensors to support sustainable agriculture and food safety, TrAC Trends in Analytical Chemistry. 128 (2020) 115906. https://doi.org/10.1016/j.trac.2020.115906

[76] A. Akbar, M.B. Sadiq, Zinc oxide nanomaterials as antimicrobial agents for food applications, in: Zinc-Based Nanostructures for Environmental and Agricultural Applications, Elsevier, 2021: pp. 167-180. https://doi.org/10.1016/B978-0-12-822836-4.00012-4

[77] M. Lv, Y. Liu, J. Geng, X. Kou, Z. Xin, D. Yang, Engineering nanomaterials-based biosensors for food safety detection, Biosensors and Bioelectronics. 106 (2018) 122-128. https://doi.org/10.1016/j.bios.2018.01.049

[78] R. Sinha, M. Ganesana, S. Andreescu, L. Stanciu, AChE biosensor based on zinc oxide sol-gel for the detection of pesticides, Analytica Chimica Acta. 661 (2010) 195-199. https://doi.org/10.1016/j.aca.2009.12.020

[79] D.X. Nie, G.Y. Shi, Y.Y. Yu, Fe3O4 Magnetic Nanoparticles as Peroxidase Mimetics Used in Colorimetric Determination of 2,4-Dinitrotoluene, Fenxi Huaxue/ Chinese Journal of Analytical Chemistry. 44 (2016) 179-185. https://doi.org/10.1016/S1872-2040(16)60902-7

[80] S. Wang, A Comparative study of Fenton and Fenton-like reaction kinetics in decolourisation of wastewater, Dyes and Pigments. 76 (2008) 714-720. https://doi.org/10.1016/j.dyepig.2007.01.012

[81] F. Garkani Nejad, S. Tajik, H. Beitollahi, I. Sheikhshoaie, Magnetic nanomaterials based electrochemical (bio)sensors for food analysis, Talanta. 228 (2021) 122075. https://doi.org/10.1016/j.talanta.2020.122075

[82] Y.T.H. Dang, S. Gangadoo, P. Rajapaksha, V.K. Truong, D. Cozzolino, J. Chapman, Biosensors in Food Traceability and Quality, Comprehensive Foodomics. (2021) 308-321. https://doi.org/10.1016/B978-0-08-100596-5.22853-9

[83] A. Lateef, R. Nazir, Application of Magnetic Nanomaterials for Water Treatment, in: Research Anthology on Synthesis, Characterization, and Applications of Nanomaterials, IGI Global, 2021: pp. 1211-1229. https://doi.org/10.4018/978-1-7998-8591-7.ch050

[84] K. Jerath, Magnetic Nanomaterials and Their Use in Water Treatment, in: 2018: pp. 109-115. https://doi.org/10.1007/978-3-319-71327-4_13

[85] A.G. Leonel, A.A.P. Mansur, H.S. Mansur, Advanced Functional Nanostructures based on Magnetic Iron Oxide Nanomaterials for Water Remediation: A Review, Water Research. 190 (2021) 116693. https://doi.org/10.1016/j.watres.2020.116693

[86] N. Goswami, A. Giri, M.S. Bootharaju, P.L. Xavier, T. Pradeep, S.K. Pal, Copper quantum clusters in protein matrix: Potential sensor of Pb 2+ ion, Analytical Chemistry. 83 (2011) 9676-9680. https://doi.org/10.1021/ac202610e

[87] L. Hu, Y. Yuan, L. Zhang, J. Zhao, S. Majeed, G. Xu, Copper nanoclusters as peroxidase mimetics and their applications to H2O2 and glucose detection, Analytica Chimica Acta. 762 (2013) 83-86. https://doi.org/10.1016/j.aca.2012.11.056

[88] W.J. Zhang, S.G. Liu, L. Han, Y. Ling, L.L. Liao, S. Mo, H.Q. Luo, N.B. Li, Copper nanoclusters with strong fluorescence emission as a sensing platform for sensitive and selective detection of picric acid, Analytical Methods. 10 (2018) 4251-4256. https://doi.org/10.1039/C8AY01357H

[89] D. Li, Z. Chen, Z. Wan, T. Yang, H. Wang, X. Mei, One-pot development of water-soluble copper nanoclusters with red emission and aggregation induced fluorescence enhancement, RSC Advances. 6 (2016) 34090-34095. https://doi.org/10.1039/C6RA01499B

[90] J.R. Bhamore, B. Deshmukh, V. Haran, S. Jha, R.K. Singhal, N. Lenka, S.K. Kailasa, Z.V.P. Murthy, One-step eco-friendly approach for the fabrication of synergistically engineered fluorescent copper nanoclusters: Sensing of Hg2+ ion and cellular uptake and bioimaging properties, New Journal of Chemistry. 42 (2018) 1510-1520. https://doi.org/10.1039/C7NJ04031H

[91] G. Zhang, T. Xu, H. Du, Y. Qiao, X. Guo, L. Shi, Y. Zhang, S. Shuang, C. Dong, H. Ma, A reversible fluorescent pH-sensing system based on the one-pot synthesis of natural silk fibroin-capped copper nanoclusters, Journal of Materials Chemistry C. 4 (2016) 3540-3545. https://doi.org/10.1039/C6TC00314A

[92] Q. Du, X. Zhang, H. Cao, Y. Huang, Polydopamine coated copper nanoclusters with aggregation-induced emission for fluorometric determination of phosphate ion and acid phosphatase activity, Microchimica Acta. 187 (2020). https://doi.org/10.1007/s00604-020-04335-2

[93] X. Yang, Y. Feng, S. Zhu, Y. Luo, Y. Zhuo, Y. Dou, One-step synthesis and applications of fluorescent Cu nanoclusters stabilized by l-cysteine in aqueous solution, Analytica Chimica Acta. 847 (2014) 49-54. https://doi.org/10.1016/j.aca.2014.07.019

[94] B. Liu, J. Liu, Surface modification of nanozymes, Nano Research. 10 (2017) 1125-1148. https://doi.org/10.1007/s12274-017-1426-5

[95] M. Zhao, J. Huang, Y. Zhou, X. Pan, H. He, Z. Ye, X. Pan, Controlled synthesis of spinel ZnFe2O4 decorated ZnO heterostructures as peroxidase mimetics for enhanced colorimetric biosensing, Chemical Communications. 49 (2013) 7656-7658. https://doi.org/10.1039/c3cc43154a

[96] A. Centeno, R. Maggi, B. Delmon, Use of noble metals in hydrodeoxygenation reactions, Studies in Surface Science and Catalysis. 127 (1999) 77-84. https://doi.org/10.1016/S0167-2991(99)80395-4

[97] N. Nath, A. Chilkoti, Noble Metal Nanoparticle Biosensors, in: Radiative Decay Engineering, Springer US, 2007: pp. 353-380. https://doi.org/10.1007/0-387-27617-3_12

[98] X. Zhao, H. Zhao, L. Yan, N. Li, J. Shi, C. Jiang, Recent Developments in Detection Using Noble Metal Nanoparticles, Critical Reviews in Analytical Chemistry. 50 (2020) 97-110. https://doi.org/10.1080/10408347.2019.1576496

Chapter 3

Use of Nanomaterials-Based Enzymes in the Food Industry

Ijaz Hussain[1], Sadaf Ul Hassan[2], Zulfiqar Ali Khan[1], Matloob Ahmad[1], Tauqir A. Sherazi[3], Syed Ali Raza Naqvi[1,*]

[1]Department of Chemistry, Government college university Faisalabad-38000, Pakistan

[2]Department of Chemistry, School of Sciences, University of Management and Technology, Lahore Campus, Pakistan

[3]2Department of Chemistry, COMSATS University Islamabad, Abbottabad Campus, Abbottabad, Pakistan

* drarnaqvi@gmail.com

Abstract

Natural enzymes perform pivotal role in all biological reactions in living things. But their practical operations are restricted due to difficulty in synthesis, reprocessing, cost, and easy denaturation. To combat these hurdles, blistering exertion is dedicated for improving these enzymes to other enzymes known "artificial enzymes." The man-made enzymes, which possess enzyme mimicking properties, have fascinated researchers' attentions. From last decade, nanozymes have attained tremendous progression. Nanomaterials-based enzyme elucidates expressive features like distinct preparative protocols, low cost, long duration for storage, and high stability towards environment than natural enzymes. This draft carries survey on 1) nanozymes literature, which is considerably explored by a diverse class of nanocomposites such as composites of halogens, carbon-based nanostructured materials etc.; 2) the recent progresses made in the fabrication of nanozymes for enzyme mimicking activity; 3) the mechanism of action, schemes to increase enzymatic activities, catalytic property and recent trends of using nanomaterials-based enzymes in the food industries.

Keywords

Nanozymes, Nanoparticles, Enzymes, Food Processing, Food Industry

Contents

1.   Introduction

Enzymes are the biological catalysts which perform a vital role in living things. These enzymes alter reactions rate with extraordinary specificity and efficiency at moderate conditions such as ambient pressure, room temperature, aqueous solutions, etc. Many applications of these enzymes have been widely known in biological reactions. For example, natural enzymes have been extensively explored in food and clinical sector [1]. Typically, enzymes require optimum temperature (30-45°C) and pH range from 4 - 7.5, for their function. Enzymes are made of proteins which mainly consist of amino acid subunits. The activity of the enzymes mainly depends on the stability of the hydrogen bonding of amino acid units with other amino acids in tertiary structure of enzymes protein. The breakage of H-bonds in biological system or in-vitro environment cannot stay intact at or above 45°C which cause the loss of enzymatic activity. Due to the huge demand of enzymes in multiple sectors, there was utmost need to choose some alternate to take advantages of enzymes at wide range of reaction conditions. Man-made or artificially synthesized enzymes were introduced to cope the industrial production challenges. During synthesis and purification of enzymes, time and labor are the necessary elements, which restricts maximum demands for enzymes production. Recently, interesting and formidable achievements have been noticed, and many artificial enzymes have been synthesized like hemoglobin, dendrimers, cyclodextrins, porphyrins, hemin, proteins, supramolecules, etc. using automatic synthesizers [2].

Nanoscience deals with artificially arranging the atoms, molecules, macromolecules, and studying their properties. Those nanomaterials having enzymatic features are called nanozymes. In 2004, the term nanozymes was coined for the first time, means copying the natural enzymes features [3]. Recently, the sciences of bioanalytical and analytical is changed by nanomaterials. In the progress of this science, a basic role has been shown by nanostructured materials to different industrial areas, like biomedicine, power conservation, storage of power, nano-catalyst casts, food, management of waste materials, and engineering technology.

Food is one of the necessary elements for all human beings. The nutritional quality of food is partially or completely destroyed by many microorganisms like bacterium, pathogens and viruses. The crucial threat for the food safety is food-borne germs. Every year in the word, various diseases which lead to 400,000 deaths, are caused by germ's infections. It is very necessary to prevent food from microbial, chemical and pathogenic contaminations. A lot of food preservative methods are exercised to maintain the quality of food and to preserve it safely. However, all conventional food preservative techniques are time taking, laborious and require extensive care to accomplish. There is a need to develop sensitive and fast way to handle food-borne germs. The industry is trying to develop possible applications of nanoparticles in the market to get optimize production. Nanostructured materials have been in developing new ingredients and food, strengthening quality control, upgrading food safety, acting as biosensors in spoiled food [4]. Mimetics are those compounds that copy the activity of other materials, being indistinguishable to the ones found in the organism (hormones, mediators, enzymes) [5]. Mimetics and natural signaling molecules are indistinguishable according to physically and chemically. Variety of nanoparticle based enzymes have been developed and their use gaining the acceleration after each day of its advantageous results. There are many important features of mimetic enzymes; few of them are summarized in Figure 1.

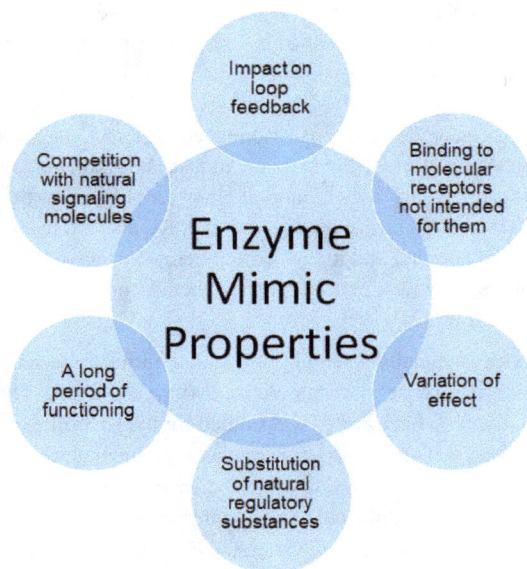

Figure 1: Features of mimetic enzymes.

2.   Nanozymes and its features

Nanozymes substitute natural regulatory substances after binding to the receptors of molecules. For the binding to receptors, there is a competition between natural signaling molecules and mimetics (by affinity and concentration). The time of linking to the receptors and the period of mimetics functioning in the same organism is not the same as for natural compounds. Because of the feedback between production of mimetics and concentration of regulators, the development of regulatory materials in the body is conditioned by the mimetics presence [6].

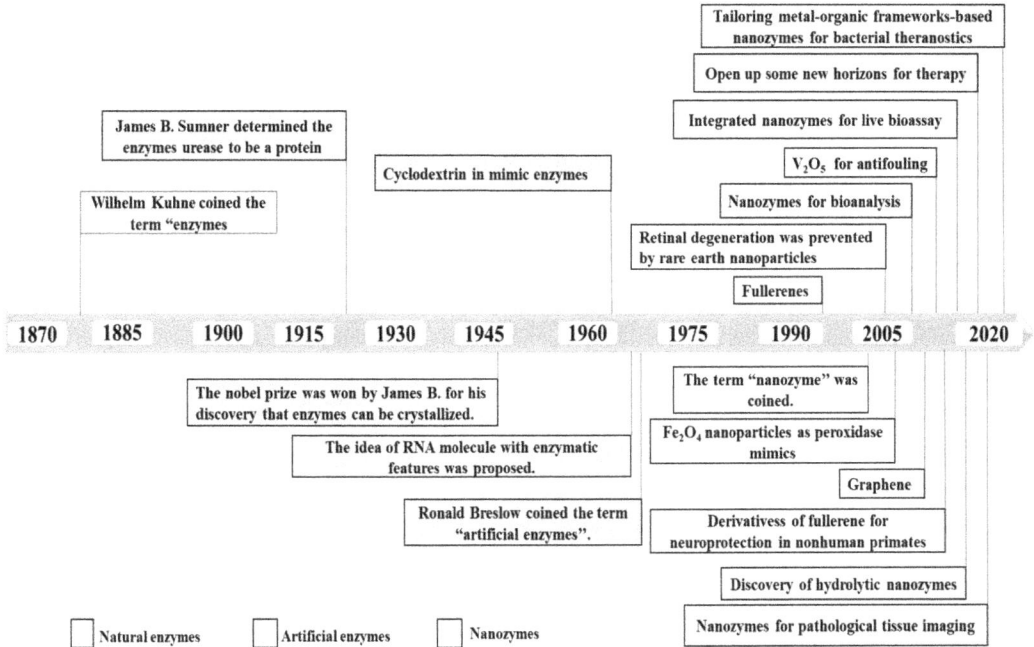

Figure 2: Complete detail for developing of man-made enzymes, natural enzymes, and nanomaterials based enzymes [1, 7, 8].

There are many advantages of nanozymes over other enzymes, however some key merits are as follow;

- The ratio of larger area of surface to volume concerning of minute size to nanozymes.
- Nanoparticles are relatively long lasting and provide longer shelf life.
- Catalytic activities of these particles are defensive via heat circumstances and a broad pH range.
- The formation of huge mass of these particles can be easily formed.

- They show pH-based properties of catalyst, structure and size which are simply attuned.

These special features of nano-structured materials convenient to modify with biological species, resulting sharp reply to exterior stimulant and leads to a superior contestant for catalysis as compared to natural enzymes. Esterase, oxidase, superoxide dismutase (SOD), peroxidase and catalase are those natural enzymes whose features are mimicked by nanomaterials [9]. Quadrametallics, trimetallics, and bimetallics are the nanocomposites which show the features similar to natural enzymes.

Recently, nanozymes have gained increasing demand and exhibit good aptitude as replacement of natural enzymes. Figure 2 shows the gradual development of man-made enzymes, natural enzymes, and nanomaterials based enzymes in detail.

Over the last few years, the research of nanozymes has achieved remarkable progress. The following sections will describe the 1) catalytic mechanism of nanomaterials-based enzymes, 2) applications of nanozymes in food industry, 3) schemes to improve substrate specificity of nanozymes, 4) nanozymes importance for the detection of food contaminants. For analytical studies of foodstuff and their vitality for human health, uttermost form of nanomaterials which could act as enzymes are peroxidase, catalase as well as oxidase, these could be applied for detection of any kind of adulteration, toxicity as well as other illegal analytes in our food products which we will elaborate with some examples and principle of action [10].

3.   Catalytic mechanism of nanomaterials based enzymes

Tunable catalytic performance is exhibited by nanozymes. However, their pragmatic applications are confined by poor substrate specificity and low catalytic property. Poor information about the true mechanism of action of nanozymes can cause delay in upgrading the nanoparticle-based enzymes for their catalytic activities. To describe the procedure about working of nanostructured enzymes, our thinking should be cleared regarding reaction mechanism, binding attraction of nanozyme for functional arbitrators, functional sites of nanozyme, substrates, and donation of electron including catalysts and substrates.

Peroxidase-like activity

To study catalytical activity mechanism of peroxidase mimics, the pathways to decompose H_2O_2 on metallic element's surface which possess brilliant resistance against chemical invasion were investigated using kinetics, showing properties similar to that of enzymes, having no dependence on the exterior passive aspects. For instance, H_2O_2 decomposition on Au(111) are mostly divided into two kinds under the normal condition. In an acidic environment, the decomposition of H_2O_2 produces OH* (* shows that labelled compounds become adsorbed on exterior of metal) after which O* and H_2O*. In this situation, O* demonstrates high oxidizing capacity to extract H atom towards organic substrates, exhibiting a peroxidase-like activity. Hydrogen peroxide can generate water as well as oxygen in the system having basic media, demonstrating property similar to catalase enzyme as compared to the pre-adsorbed OH group. This ending shows various trajectory for decay of H_2O_2 could activate the properties similar to enzyme [11].

Catalase-like activity

Majority of the nanoparticles possess catalytical activity similar to this type. The principle of action in these nanomaterials based enzymes depends upon the catalytical conversion of hydrogen peroxide into water as well as oxygen molecules. The progress of catalytical activity in these materials also based upon the type of structure of the nanoparticles which they possess during performance. For example operational procedure of cerium nanoparticles has been determined which possess enzymatic activity similar to that of catalase and also similar in catalytical activity as of superoxide dismutase. The action of working includes oxidation reduction reactions which takes place among cerium ions having oxidation states of +3 and +4 and also on availability of oxygen. Bond among oxygen atoms in hydrogen peroxide segregate which is present on exterior surface due to encouragement from protons, which results in the removal of water molecule from exterior [12].

Oxidase-like activity

The types of nanoparticles which possess catalytical activity similar to that of oxidase are used for determining the amount of analyte of interest as well as their identification, whose detection is based upon type of analyte material which could accelerate reaction rate in the presence of oxygen molecule and results in the generation of oxidized species as well as hydrogen peroxide. The progress of activity is also linked with exterior surface of analyte as well as surface of nanoparticles [13, 14].

4.   Nanomaterials-based enzymes for food analysis

4.1   Metal oxide-based

Iron oxide (Fe_3O_4) has exhibited remarkable uses in biotechnologies along with biosensing, between the high oxide value of metal-dependent nanostructured materials. Different sizes of Fe_3O_4 nanoparticles exhibited deep-rooted copies of peroxidase, that can promote the oxidization of diazoaminobenzene (DAB), o-phenylenediamine (OPD), as well as TMB, granting brown, orange, along with blue products [2]. Experimentally, the lower interaction of HRP to TMB, is because of higher M. Menten constant (Km) value of Fe_3O_4 nanocrystals. The presence of ferrous ions supports the peroxidase-alike properties of nanoparticles of Fe_3O_4. Additionally, nanocomposites of Fe_3O_4 having various nanostructured materials of Fe_3O_4 can be described to have deep-rooted oxidase along with peroxidase-like properties, showing wide implementations in medical diagnosis, therapy, environmental science, and biosensing.

The oxidase, SOD, catalase, and phosphatase mimetic activities of nanoparticles of CeO_2 are the most broadly used metal-oxide-dependent copies of enzyme. Polyacrylic-acid and dextran- coated CeO_2 nanocrystals, having deep-rooted oxidase-viz properties in the surroundings of acidic medium have been developed, that could change the oxidation of tiny species without any kinds of oxidizing agents and a serial of organic dyes as well. Population of world is increasing, consequently the demand of food is also increased. For better yield and meeting the demand of food almost 1.89 billion of people utilized toxic insecticides or pesticides to shield crops from pests. Because of excessive utilization of these insecticides around the world about 25 million people suffer from toxicity of these pesticides in every year [15]. Ethoprophos which is a pesticide as well as nematicide along with nanoparticles of cerium was developed in dual scheme as catalyst for the identification of organophosphorus, in this system nanomaterials which act as catalyst promote organophosphorus conversion into nitro-phenolic which cause appearance of yellow color which was directly related with organophosphorus concentration [16].

Antioxidants are bioactive components which are present excessive in daily vegetables as well as fruits. These are mighty weapon for our body competing to heart diseases. These antioxidants may involve phenolic compounds, ascorbic acid as well as gallic acid or some other tocopherols etc. Consequently, assessment of antioxidant activity of food products containing these compounds is of quite importance. The assessment of antioxidant ability of food products with the help of nanomaterials is mostly developed on antioxidant based utilization of oxidational products as a result of catalytical activity. Jia and his colleagues [17] examined antioxidant abilities of natural antioxidants like ascorbic acid, tannic acid as well as gallic acid whose working principle was based upon inhibition effect of oxidized ABTS produced as a result of catalytical activity similar to that of peroxidase like of oxide of cobalt nanoparticles and they possess different progress towards antioxidant activity and there order of activity was from tannic acid at first and ascorbic acid at last. On the other hand, Jin and his colleagues [18] developed a quite different mechanism of action for ascorbic acid identification which was dependent upon ascorbic acid stimulated deterioration of nanomaterial which acts as an enzyme. In this method ascorbic acid can untangle nanostructures of COOH similar to that of oxidase into cobalt ion therefore it could inhibit catalytical activity of tetramethylbenzidine. This detection system possesses an excellent response towards ascorbic acid having limit of detection up to 140 nM as well as very small duration for assaying of about 5 minutes which could also be utilized for sample of orange juice.

For proper growth and nourishment of brain as well as development of memory food enriched in acetocholine are quite helpful, because signalizing of choline play a pivotal role in the nervous system of the brain. It is quite important to fabricate a simple as well as loyal procedure for determination of choline in foodstuff. Zhou with his colleagues [19] fabricated a normal colorimetric procedure for determination of acetocholine or choline by using tandem reaction with the help of iron oxide and re-oxidized graphene oxide nanomaterials like acetocholinesterase inhibitor as well as choline oxidase which possess catalytical activity similar to peroxidase. This approach can detect 10-100 mM of range for choline and possess limit of detection of about 38 nM. In real sense, analytical study of samples of milk elaborate its potency in foodstuff investigation applications. Likewise, Nirala and Prakash [20] performed a fast colorimetric determination of acetocholine in samples containing milk by application of molybdenum disulfide nanostructures with peroxidase activity as well as choline oxidase. With the help of magnetic nanoparticles, melamine was determined in samples of milk having detection limit up to 2.6 ppm with optimized conditions which was about at pH of 4, temperature was about 46 °C and time of assaying was around 10 minutes. By inspiring from these unique characteristics of magnetic nanoparticles, researchers also worked on enzymatic activity of oxides of iron, copper as well as oxides of Mn [21].

Ions of Mn replaced with oxide of cobalt having submicron spherical structures was linked with antiochratoxin peptide to copy oxidase like activity and for the determination of ochratoxin in corn samples. This form of enzymatic material exhibit 50 times better attraction against ochratoxin as compare to that of pure enzyme. This increment in concentration of ochratoxin cause rise in oxidation of tertrametylbenzidine, which show change in color at wavelength of about 652 nm having limit of detection about 0.079 ng/ml [22].

4.2 Metal-based nanozymes

Due to their attractive electronic features and good biocompatibility, metal-based nanomaterials are of considerable importance in nanotechnology, resulting broad implementation in sensing, biological medicine, electrolyser, electrocatalysis, biological medicine, imaging, and catalysis. Metal nanomaterials copies the property of oxidase of glucose, catalase, peroxidase, SOD, oxidase, because of having intrinsic enzyme-like features. Nanocomposites of silver are most broadly known, among a large class of metal nanostructured materials. The surface property of nanocrystals of Au influence the property of peroxidase. Negatively charged silver nanocrystals having electronegative character, altered by any salt of citric acid, show lesser copying of peroxidase than those by prepared positively charged Au nanoparticles. A lot of research works have been done to judge the ligands effect on enzyme-like catalytic activity. In addition, enzyme-like efficiency is determined by pH values [2].

Silver nanocrystals normally exhibit the higher properties similar to peroxidase (around pH = 4.0) in an acidic solution, but it is different about physical and biological values of pH, limiting the biologic implementation. By using Lipo-Hepin at normal circumstances, the peroxidase-alike property of nanocomposites of silver has been enhanced [23]. All the kinds of copies of enzyme have been managed by the accumulation of ions of metal on silver nanocrystals surface.

Whereas an essential component for the human body is glucose which is associated with diagnosis or worse feeling of diabetes. So precise as well as exact determination of level of sugar or glucose in our foodstuff is quite necessary which could be beneficial for guiding and consulting about diet for patients with diabetes. For determining the level of glucose in foodstuff sample a new sensing system has been fabricated which working principle is related to nanozymes, having enzymatic activity similar to that of peroxidase as well as oxidase. Huang and colleagues [24] developed an assay for glucose quantification with the help of cascade reaction of GOX with nanozymes having catalytical activity similar to peroxidase. Under controlled situations this assay analyzed a varied range of concentration from 5-250uM but this assay has comparably low detection limit of about 1.49uM which can be utilized fairly for samples of juice [21].

In an attracting enzyme pounding investigation, effect of ions of halogen family on a protein casein with modified nanoparticles of gold and their catalytical activity was elaborated. In the whole halogen family iodide ion possess prominent as well as fast irreversible inhibition effect of nanomaterials for peroxidase like activity and indicate the reason due to bonding among composite materials also cause blockage of functional centers of enzyme. The assembly also indicate that halides pounding capability was based upon exterior surface as well as intrinsic characteristics of modified materials applied [10]. Because an indigenous constituent of wine as well as ethyl alcohol level is necessary to be analyzed regarding defilement assessment of these alcoholic liquors and concerning dietary admonishments. Stasyuk and his colleagues [25] developed a detection system whose working principle was based upon electrical changes by connecting nanoparticles of platinum and ruthium with oxidase alcohol for inspection of primary alcohols in alcoholic liquor. The attachment of pure peroxidase with composites of platinum and ruthium cause development of sensing system which act as artificial enzyme like catalyst and enhance electrochemical abilities of working electrodes as well as increment in linear range.

Sensitivity of this approach is quite excellent which is 335 ampere/$M^{-1}m^{-2}$ for ethyl alcohol and range of 20-200 uM, as well as better selectivity for pure analytes.

One of most customary issue faced during protection of foodstuff is food allergens which must have more precise as well as sensitive approaches of analysis for determination of these allergens. Latterly, a new trending NLISA methodology was developed for uttermost sensitive resolution of the allergens in milk samples of cow with the application of β-lactoglobulin as biological measure for proteins of milk. In contrast to this assay, writer ambition was to enhance limit of detection by applying platinum nanoparticles which possess peroxidase like catalytical activity mounted with different antibodies and horseradish peroxidase molecule as new technique for signaling [10].

4.3   Metal-organic frameworks based nanozymes

Recently, many types of metal-organic frameworks (MOF) can be shown to have peroxidase, laccase, as well as oxidase viz property. Furthermore, Cu based metal organic frameworks which can be produced with the help of 2-aminoterephthalic acid along with copper nitrate and in control of various mechanism, exhibiting property similar to peroxidase, which can produce yellowish colour by catalyzing oxidation of TMB along with the help of hydrogen peroxide [26].

The developed Ce-based MOFs showed outstanding property similar to oxidase due to unforced reprocessing along with flip-flop in the oxidation and reduction surrounding of Ce^{3+}/Ce^{4+}. Furthermore, these nanomaterials-based frameworks show lesser value of Michaelis constant as well as greater starting value of velocity (Vmax) after comparing with CeO_2, resulting greater interaction and TMB catalytic property. In addition, Ce-MOFs peroxidase-viz property have been assigned for greater plane along with p–p assembling interactivity in TMB and nanostructured-based frameworks. An excellent strength is exhibited by Ce-MOFs, resulting good implementations in the biothiols diplomatic finding.

Metal organic framework, nanocomposite type of metal organic framework as well as modified form of these materials can be applied comfortably for various detection systems for monitoring foodstuff products because of their proficient nature. MOFs in original form possess relatively enhanced porous area in their original structure, as well as resilient functional centers which cause excellent catalytical activity of biological molecules. Many forms of organic based framework for instance iron based MOF in combination with $88NH_2$ possess brilliant catalytical activity similar to peroxidase which is helpful in determination of hydrogen peroxide in sample of milk, belong to class of MOF in their original form. When these are applied with luminol they show excellent catalytical activity of reaction and express enhanced chemiluminescence signal and detection of hydrogen peroxide was quite precise. The limit of detection was lying around 0.025umol/L [27].

Unsaturated functioning group coordinately attached across metal organic framework promote alteration of functioning sites, which cause promotion of functional ability of these nanomaterials. For example in determination of acetylocholine or choline in samples of milk. Valekar with his friends [28] developed MOF based nanocomposite with simple diamines across coordinately unsaturated functioning center and determined enzymatic activity similar to that of peroxidase. The maximum catalytical performance was observed due to symbiotic effect of boosted negative efficacy, as well as accurately establishing the diamines molecular size. Chen with colleagues [29] developed PtNPs/MOF composite formed of Cu, made of copper, as of iron containing prosthetic group (Cu-TCPP(Fe)) ligand, and integral platinum nanoparticles were applied in determination of hydrogen peroxide. This type of composite materials possess brilliant catalytical performance having limit of detection of about 0.36uM in contrast to pure constituents which developed metal organic based framework composite as well as prevented accumulation of nanoparticles of platinum on very thin nano-disk during development [10].

Hydrogen peroxide is mostly chosen preservative as well as bactericide for foodstuff containing raw milk samples for the protection as well as keeping milk sample fresh. Moreover, extreme level of hydrogen peroxide to sample of milk not only promote deterioration of essential components of milk but also cause induction of intestinal as well as nervous disorders to the consumer. According to world organization of food and agriculture, the maximum level of acceptance of hydrogen peroxide is up to 0.05% (w/w). Freshly, different types of methods have been introduced for precise detection of hydrogen peroxide in dairy foodstuff, and most of sensing system for detection of hydrogen peroxide are colorimetric assay. Liu with his colleagues [30] introduced colorimetric system for determination of hydrogen peroxide with the help of iron doped copper tin hydroxide nanozymes having similar catalytical activity as of peroxidase and the substrate was TMB and range of working was from 30 to 1000 uM and limit of detection was around the 9.5 uM. But limit of detections of these assays are not better enough up to satisfaction level which demands for some more novel sensing systems like as fluorescence detector.

4.4 Molecularly imprinted polymers (MIP)-Based

Molecularly imprinted based nanopolymers having enzymatic activity are developed by the phenomenon of polymerization across the main element. This procedure generates active centers to the main surface, which can be renewed and recycled inattentive of their saving circumstances, and consequently examined an outstanding substitute for natural sense-organs [31]. To improve the peroxide reductase activity of Fe_3O_4 nanomaterials, their plane was stamped by TMB rigorous the adaptation to zeta potential. This developed in the evolution to patches of substratum irrevocable, and a hundred-fold enhance to particularly was noticed after contrasting with non-stamped nanostructured-based enzymes [32]. Many researchers toiled on increasing the glucose oxidoreductase copying the property of nanocomposites of Au/Pt by elemental imprinting procedure. They exhibited higher property by using the ratio (1:1) and were amalgamated along with magnetic nanospheres for uncomplicated dissociation and good strength. Glucose-irrevocable amino-phenyl boronic acid was utilized for printing after the combination of main element of glucose. The patches of glucose irrevocable were synthesized after the elimination to main elements, and showing two hundreds-fold greater catalytical regulation as compare to nanoparticles of Au [33]. After the study of patulin existing in the juices of fruit, MIPs remodeled nanoparticles of Au, chitosan, and C dots are applied. To use of chitosan along with CDs enhanced the shifting rate of electrons, the plane area of electrode, and power of signal, that are identified. The limit of detection was about to $(7.58\times10^{-13}$ mol/L) [34].

4.5   Carbon-based nanozymes

Nanomaterials of carbon-based like (C-dots), grapheme (GE), fullerenes, as well as nanowires of carbon generally show remarkable physiochemical features, that are performing a vital role in the fields of research; such as catalysis, engineering of surroundings, transformation, biological medicine, electronics and optoelectronics, power storehouse, and sensing. In the last few decades, deep-rooted peroxidase-like property is exhibited by carbon nanostructured materials. It is noticed that modification of surface along with hydroxyl as well as carboxyl classes imparts remarkable character as peroxidase-like activity. Qu with colleagues [35] demonstrated peroxidase-like property of oxide of carboxylic reshaped grapheme for the first time. After further investigation the oxide of carboxylic reshaped grapheme demonstrates greater interaction to the enzyme acting place 3, 30, 5, 50-tetramethylbenzidine (TMB) as compare to enzyme horseradish peroxidase (HRP), having good activity of catalyst. Likewise, the peroxidase-like property is also shown by carboxyl-reshaped C_{60} ($C_{60}[C(COOH)_2]_2$) [36].

Due to unique physical and chemical features, C-dots have aroused great attention since 2004 [37]. For example, C-dots can also be seen to show the property similar to peroxidase, by receiving the lone-pair of electrons to H_2O_2 from TMB amino groups, resulting to electrons shift from C-dots. Some other kinds of nanocomposites such as SWCNHs can also exhibit an efficient peroxidase-like property. In addition, the greater stability against pH and temperature have been exhibited by l-the prepared carboxylic-group-functionalized SWCNHs than those by HRP [2].

The nanocomposites of Pd-C dots are applied to identify glucose and H_2O_2 in direct specimens, and limit of detection announced for H_2O_2 about (0.3 μM) are mainly same to the the value of LOD prevailed. The composites of carbon dots and metal-based nanoparticles, provide fast identification to general foodstuffs microorganism *Staphylococcus aureus*. This procedure including the nanoparticles of silver-dependent attached to microorganisms functioning centers and established sandwich immunocomplex, assisting the o-phenylenediamine (OPD) oxidation. In that event, the dose of microorganisms are directly related to the colour changes, and limit of detection was seen to be (4.91 CFU/mL) [38]. In the glucose identification, the hybrid grapheme nanocrystals of N_2 and Ni along with nanoparticles of Pt showed peroxide reductase-viz property. This catalytical nanocomposite are demonstrated to beneficial for the immobilization of glucose oxidoreductases enzyme, consequently identification of the glucose quantity, utilizing micro-fluidic paper-dependent nanomaterials) [38, 39].

A lot of heavy metals causes the diseases like minimata and Plumbism in the food chain and pure water. Even in very small amount, arsenic, mercury, cadmium, and lead are those heavy metals which may cause terrible effects [40]. These heavy metals are mostly identified by nanocomposites of graphene oxide-based nanoparticles along with film of bismuth, owing to its more conductivity and limit of detection are noticed to be (0.50 μg/L) and (0.80 μg/L) for the ions of Cd and Pb. Various kinds of tea specimens are distinguished based on 12 ions of metals such as (Ni^+, Ag^+, Ba^{2+}, Cr^{2+}, Sn^{2+}, Cu^{2+}, Pb^{2+}, Mg^{2+}, Mn^{2+}, Ca^{2+}, Fe^{3+}, Al^{3+}) using 3 kinds of oxidase-copying nanoparticles-based enzymes (Cu/AMP Cu/ATP, and Cu/ADP) [41].

5.   Schemes to improve substrate specificity of nanozymes

We know that natural enzymes induce complete specificity for a specific reactant. While nanomaterials which possess enzymatic property display catalytical activity scheme that are most appreciative, but their specificity for the reactants are not so good. Many of oxidase as well as peroxidase like activity, the oxidation of TMB cannot be only catalyzed by nanozymes, but on the other hand, for other substrates like OPD and ABTS this may occur. Hence, the basic task here for us is to make innovative types of nanomaterials which can behave as enzymes with outstanding selectivity for the specific form of reactant. For the improvement of nanozymes binding selectivity for the special target material, this may be useful to raise functional binding centers of substrate by using the approach of transformation chemically to upgrade the substrate specificity of these enzymes [42]. A distinctive way has been proposed to improve the oxidize reactant specificity is to connect peptide molecule or strand to DNAzymes [43].

Furthermore, Liu with his colleagues [30] showed a new way of molecular imprinting to enhance the specificity for oxidation activity of TMB in contrast to ABTS. They have produced polymers which possess imprinted molecules located onto the exterior of Fe_3O_4 nanoparticles to examine about their specificity features for showing peroxidase properties similar as that of peroxidase. The oxidation of ABTS and TMB can be catalyzed by naked Fe_3O_4 nanoparticles, but imprinting of TMB sets on the core of Fe_3O_4 molecular imprinted polymer accelerate the catalytical property and selectivity to the oxidation of TMB, and ABTS-imprinted sets along with the core of Fe_3O_4 (A-MIP) promote the catalytical property and ABTS selectivity towards the reactant. The efficiency of a catalytical activity is applied to contrast the performance of their substrate specificity, exhibiting good selectivity possessed by imprinting with the reference of adsorbed reactant. In addition to, by applying monomers in which charge has been induced, show almost 100 times good selectivity for the imprinted reactant species as compared to the naked Fe_3O_4 nanomaterials. As described earlier in above portion, nano-sized particles coated with metal single layered can accelerate the enzyme activity of nanozymes [44].

On the other hand, there is an excellent scheme to promote the nanozymes substrate specificity with the help of the monolayer modification. For instance, differentiation of enentiomers can be attained by modification of chiral monolayer on Au nanozymes. The chiral nanomaterials which possess enzymatic activity having bimetallic catalytical centers of chiral nanozymes were manufactured through the mutual connection of thiol-having chiral ZnII-containing major groups on Au nanoparticles exterior, which could display diverse properties to various substrates of RNA dinucleotide. Hence, nanozymes exhibited outstanding selectivity for reactant for the splitting of dinucleotides such as ApA, GpG, CpC, or UpU. Mainly, a remarkable reactivity difference regarding the splitting of UpU enantiomer was noticed, exhibiting excellent capability for enantiomer differentiation [2].

6.   Some other applications in the food industry

Recently, a lot of nanozymes have been synthesized which possess enzyme like properties and have exhibited an excellent forum for routine uses. In this chapter, our main focus of discussion will be upon the routine applications of nanomaterials which show enzyme like activities, in the food industry, depending upon the catalytical mechanism similar to enzymes. Most blistering, versatile, and simple form of detectors having excellent selectivity as well as high sensitivity have already been discussed. Nanomaterials based enzymes are used in food industries for various purposes.

Table.1 *Food industry application of nanostructured based enzymes*

Nanostructured-materials	Biological conjunction Strategy	Uses	References
Nanocrystals of carbon	Fructose oxidoreductasde adsorption	Fructose identification with the limit value (.001 mM) in honey	[45]
Zinc oxide:Cobalt nanocluster	Glucose oxidoreductase covalent immobilizing	Glucose identification with the range of (0.02-0.21mM)	[46]
Nanoparticles of carbon	Lactase covalent immobilizing	*E. coli* identification with the range of (10-10^4cfu/ml)	[47]
Nanocrystals of Au	Pyranose-2-oxidase (POx) covalent immobilizing	Glucose identification with the range of 0.05-.75mM	[48]
Nanoparticles of carbon	Aflatoxin oxidoreductase covalent immobilizing	Aflatoxin identification with the range of (0.5-2ng/ml) in food	[49]
Nanocomposites of Au	Lactase absorption (Physically)	Identification of *Escherichia Coli* with range of (100 *E.Coli*/ml)	[50]
Nanomaterials of zinc oxides	Xanthine oxidoreductase covalent co-immobilising	Xanthine identification with the range of (0.1-100 μM) in fish meat	[51]
Films of zinc oxide nanorod (ZnONR).	Choline peroxide reductase and oxidoreductase covalent co-immobilising	Phosphatidylcholine identification with the range of (0.0005-2mM) in milk	[52]

6.1   Intentional adulteration

The embezzle use of illicit supplements of food has ensued of persistent exposure to success of experiences of food welfare, that has created a serious ultimatum for the well-being of public and caused general troubles for the well-being of food. Owing to very toxicity of these illicit supplements of food, the survey of illicit supplements of food is

Materials Research Forum LLC
https://doi.org/10.21741/9781644901977-3

pivotal for the well-being of food [53]. Nitrite, clenbuterol (CLB), and Sudan I are considered to be the general illicit supplements of food, that caused global threat in most countries. Various nanoparticles-based enzymes are being synthesized for the fast identification of these additives. Sudan I is being widely used as supplements of food, especially in the powder of red chilly due to attractive red and inexpensive. But, Sudan I are known to be poisonous, and can cause harm to digenetic substance because of the interaction along with in vitro particular sequence of DNA [54]. A good, reliable, and novel nanoparticles of Platinum modified electrode graphene-β-cyclodextrin are being developed to identify Sudan I in the specimens of food. The electrochemical appearances of many remodeled electrodes for this compound are being assessed by cyclic voltammetry. All of these remodeled electrodes composites, the developed nanoparticles of grapheme-β-CD-Pt altered electrode revealed more electrocatalytical property for Sudan I. Furthermore, the developed remodeled electrode can ride the susceptibility of Sudan I, and resulting to a direct retaliation range improvement of electrochemical nanomaterials. These nanoparticles can be applied to identify samples of Sudan I (chili powder, and ketchup) having range founding from (0.005 to 68.69 μM), and limit of detection (1.6 nM) [55].

Major concern in food products vitality which may be resolved by the identification as well as handling credentials of nanomaterials, are pollutants and adulterations caused by many of bacterium as well as protozons. In the same appreciation, a detection system has been developed whose working principle based upon electrochemical changes on $Ca_3(PO_4)_2$ for tyramine amino acid identification and analysis in cheese containing foodstuff by applying tyrosinase enzyme as a substance for biological system. The obtained detection system was feasible, reliable, sensitivity was extremely good, fast, cheap, and easy to handle for detection of tyramine. In feasible circumstances, a straight line limit lies between 6×10^{-7} to 1.7×10^{-5}, response signal of 1.6×10^3 mA M^{-1} cm^{-2}, detection limit was 4.85×10^{-8} M, and the time for each response was about 6s [56]. A handsome rate of flow of negatively charged electrons was obtained among the functional binding sites of enzyme because of better adhesive force of AChE with hydrophobic exterior of working electrode, that accelerate recurrence of reactant to functional sites of catalyst. The developed detection system was applied in the identification of insecticides in pointed tomato cocktails. Their obtained results reveal limit of identification of carbaryl insecticides lies from 93% to 105%, and in case of organophosphate insecticides, this occurred in the range of 91% to 110%. These values persistent that developed detection system has better determinacy as well as detection proficiency in analysis of foodstuff related issues [57].

6.2   Detection system for insecticides

Latterly, many scientists argued the preparation strategies as well as applications of a detection system whose working is associated with Pt nanoparticles type materials bonded with UiO66-NH$_2$ as facilitator to develop acetylcholinesterase detection system for insecticides of organophosphorus class, detection in food products containing apple as well as cabbage. Their observations revealed that our established detector possess highly amplified signal for identification, ranging from 10^{-14} M to 10^{-9} M, having an identification range of 5×10^{-15} M. Applying this recommended detection system, malathion insecticide limit of identification was lying in the range between 93% to 98%. These values elaborate that our detection system is a feasible choice for suspicious applications [58].

6.3   Design for detection of gram negative bacterium

In same field of studies, researchers utilized de-oxy ribozyme for the fabrication of a detection system for determination of concentration of interested components whose working was associated with a catalectic peptide and trivalent DNAzyme design for, detection of gram negative bacterium *Vibrio parahemolyticus*, known as a foodstuff-propagated mutagen which is mostly present in common food fishes. In the following studies, peptide molecules associated with nanoparticles which possess magnetism ability (MNPs) were utilized for carrying and facilitating purpose, and the G4 secondary structure which possess stable DNA configuration was applied for amplifying signal of reactant. In feasible circumstances, a reasonable linear signal range for identification was obtained from 102 to 107 CFU/mL, also having a lesser identification limit of about 10 CFU/mL. Consequently, our developed detection system possess better as well as amplified signal, ensuring it a feasible choice for the assurance of foodstuff vitality instead of utilizing complex procedures [58].

6.4   Detection of ethanol

Detection of ethanol is vital in clinical analysis and also in beverage production industries to monitor the progress of fermentation processes. Many types of biosensors has been developed by many architectures for related studies. The founder utilized nanoparticles of octahedral isotope of carbon which was modified by chemical interaction with C$_{12}$H$_9$NS which act as conveyer for dehydrogenase enzyme which promote the oxidation of ethanol. The detection system developed show a certain catalytical activity in electrical system and express oxidation activity of NADH, having better discrimination for ethyl alcohol, and interpretation was enjoyable [59]. Consequently, many kinds of detection system whose working is associated with nanostructured materials might be suggested for several ambitions in field of food products, planning for better identification of insecticides as well

Materials Research Foundations 126 (2022) 89-116   https://doi.org/10.21741/9781644901977-3

as protozons or bacterium, as illustrated earlier. It is legible prospect of detection system in the field of technologies which is developing day by day to guarantee best regulation, screening, and vitality of foodstuff, by keeping in mind the facts in that area, which is directly related with the betterment and dignity of diverse community or environment. That's why, there is much need about feasible, tenable regime as well as expedite for degradation of challenges which influence on environment by facilitating recent advancement, which determine the obsession as well as level of pathogenic, which could deteriorate foodstuffs, sludge as well as plants has developed an area of interest and charm in research which is planned for certifying the manufacturing as well as conservation of intact and vital foodstuffs [60].

6.5   Mycotoxins

Mycotoxins are very noxious for living things and cause long-lasting protection peril towards the products of agri-food, which is very poor as contrast to illicit additive. The development of tactful and particular procedures are motivated by the powerful negative results of mycotoxins. Huang with his colleagues [61] manufactured a nanozyme-dependent colorimetrical aptasensor of $MnCo_2O_4$ for the detection of biomolecules, and the OXD-viz property of $MnCo_2O_4$ nanomaterial are being controlled by aptamer reversible loyalty on a nanosphere plane of $MnCo_2O_4$ umpired by target identification. It is revealed that the ochratoxin is sensitively, and quickly identified in the sample of maize, along with value of LOD (0.08 ng/mL). Mainly, the concurrent identification of many selected elements is more desirable as compare to identification of the single element. [62] described the micromotor-dependent "on-the-fly" incandescent perspective by using composite of catalase-viz, rGO-Pt nanoparticles, that provided fast and concurrent identification FB1 (0.4 ng/mL) and OTA (7 ng/mL), corresponding greater perceptivity. During the investigation of the specimens of wine and beer, an important quantitative recuperations (97%) were acquired [21].

6.6   Other food contaminants detection

Additionally, many defilements of food are identified by nanomaterials-based enzymes, like norovirus (NoV), lipopolysaccharide (LPS), hydroquinone (H2Q), and arsenic (III) [54].

6.6.1 Lipopolysaccharide (LPS)

Shen with his colleagues [63] confirmed a nanomaterial of ratiometric having greater perceptivity and validity to identify LPS with the help of nanocomposites of Cu-MOFs for the exaggeration of signal.

6.6.2 Hydroquinone (H₂Q)

For the first time, the cerium vanadium oxide (CeVO₄) was developed by using very easy scheme, displaying both oxidoreductases along with peroxide reductase-copying property. CeVO₄ can also be preferred for catalysing TMB oxidation to build a blue colour with the presence or without the presence of hydrogen peroxide. H₂Q can be lowered to give rise to a disparity of visible colour while CeVO₄ oxidizes TMB. But, the isomer of H₂Q (CC and RC) can't. Correspondingly, a colorimetrical nanomaterial was built for discrimination of H₂Q from RC along with CC, and H₂Q can be identified in direct retaliation having range of (0.05-8 μM) along with value of LOD (0.04 μM) [64].

6.6.3 Arsenic-III

The formation of new hypersensitive scheme to identify the Arsenic (III) is very advantageous and remarkable owing to low concentration and high toxicity of it in drinking water. Lately, Li with colleagues [41] synthesized the tactful discerning interface to identify As (III) with the help of the developed dullard-viz nanoparticles of Au-Fe₃O₄. According to the amalgamation of silver nanoparticles catalysts, the surface-active conciliation of iron (II), Fe₃O₄ nanoparticles adsorption, and the electrochemically retaliation to identify As (III) was adequately magnified. The nanoparticles of Au- Fe₃O₄ can be altered to C electrode having screen-stamped for the production of discerning interfaces.

6.6.4 Norovirus (NoV)

Owing to the demand of elimination to inhibitory elements and virus matrix particular concentration, the norovirus (NoV) identification in the samples of food is highly exigent. Lately, Weeranthunge with his colleagues [65] synthesized a new, hypersensitive, along with extremely powerful colorimetrical nanomaterials to identify murine norovirus (MNV). The nanoparticles of silver along with the peroxide reductase-viz property known as nanostructured enzymes to catalyze the colorless substratum of TMB to give rise to the product of a blue color. But, these norovirus-particular AG3 aptamer elements can hinder the peroxide reductase-viz property of nanomaterials-based enzymes owing to their absorption over the AuNPs surface. With the existence of MNV, the MNV-particular aptamers of AG3 are eliminated from plane of the AuNPs in the presence of MNV because of particular binding to MNV, appearing in the catalytical property recuperation of nanoparticles of Au, once more giving rise to the product of blue color. The quantity of MNV in specimens is in proportion to the color strength change. While, these aptamers do not have sympathy to other defilements, so the catalytical property of nanoparticles of Au cannot be recuperated and no color alter with the presence of general target. With the amalgamation among the peroxide reductase-viz property of nanoparticles of Au. By the developed colorimetrical nanoparticle, the MNV was identified to LOD for twenty viruses/ assessment near to two hundreds viruses/mL [66].

Conclusion

In this chapter novel advancements in, catalytical features, synthesis, as well as uses of nanomaterials based enzymes has been elaborated. Many types of nanomaterial based enzymes have been explained, also some innovative schemes which could enhance catalytical property as well as discrimination have been mentioned. Nanomaterials assumed enzyme are now being synthesized to use in many interesting areas. The approaches in nanotechnology incorporation with the already existing nanomaterials which possess enzyme like properties, has also been showing enhanced suitability of nanomaterials. This could be observed that most of nanomaterials are among the most impressive materials; that could increase the transducers signal as a result of electrochemical changes, decrease retort time period, and are biocompatible as well as strong to other analogous non-nanomaterials.

For the safety of food, most of the nanoparticles depend upon colorimetric assay. There are many drawbacks of this assay. Therefore, the combinations of various assays are preferred like approach of dual colorimetric-digital. This may help to future directions for the development of nanostructured-based enzymes for food industry. We should not use nanozymes only in the areas of food safety, nanozyme can be used for other resolution areas, for instance, transcription of DNA, catalytical activity of enzymes, transmission of nerve signal, and transduction of hormone signal. But, prior to attaining those objects in the upcoming days, scientists should give heed to the nanomaterials-based enzymes itself bioavailability as well as biosafety. In spite of, many accessible research on the properties of nanomaterials-based enzymes as well as applications, contrasted to the conventional enzymes, those works are yet rather restricted. So, large examinations in the upcoming days should be performed to fix problems as well as investigate further utilization of these enzymes.

Acknowledgment

The authors are thankful to Higher Education Commission (HEC), Islamabad, Pakistan and Government College University Faisalabad for providing resources to complete this project.

References

[1] X. Wang, W. Guo, Y. Hu, J. Wu, H. Wei, Nanozymes: next wave of artificial enzymes, Springer, 2016. https://doi.org/10.1007/978-3-662-53068-9

[2] W. Song, B. Zhao, C. Wang, Y. Ozaki, X. Lu, Functional nanomaterials with unique enzyme-like characteristics for sensing applications, Journal of Materials Chemistry B. 7 (2019) 850-875. https://doi.org/10.1039/C8TB02878H

[3] F. Manea, F.B. Houillon, L. Pasquato, P. Scrimin, Nanozymes: Gold-nanoparticle-based transphosphorylation catalysts, Angewandte Chemie. 116 (2004) 6291-6295. https://doi.org/10.1002/ange.200460649

[4] V.A. Kumar, T. Uchida, T. Mizuki, Y. Nakajima, Y. Katsube, T. Hanajiri, T. Maekawa, Synthesis of nanoparticles composed of silver and silver chloride for a plasmonic photocatalyst using an extract from a weed Solidago altissima (goldenrod), Advances in Natural Sciences: Nanoscience and Nanotechnology. 7 (2016) 015002. https://doi.org/10.1088/2043-6262/7/1/015002

[5] S. Singh, Cerium oxide based nanozymes: Redox phenomenon at biointerfaces, Biointerphases. 11 (2016) 04B202. https://doi.org/10.1116/1.4966535

[6] I. Chekman, N. Horchakova, P. Simonov, Biologically active substances as nanostructures: A biochemical aspect, Clinical pharmacy. 21 (2017) 15-22. https://doi.org/10.24959/cphj.17.1422

[7] S. Li, Z. Zhou, Z. Tie, B. Wang, M. Ye, L. Du, R. Cui, W. Liu, C. Wan, Q. Liu, Data-informed discovery of hydrolytic nanozymes, bioRxiv. (2020). https://doi.org/10.1101/2020.12.08.416305

[8] S. Ali, A. Hendriks, R. van Dalen, T. Bruyning, N. Meeuwenoord, H.S. Overkleeft, D.V. Filippov, G.A. van der Marel, N.M. van Sorge, J.D. Codée, (Automated) Synthesis of Well-defined Staphylococcus Aureus Wall Teichoic Acid Fragments, Chemistry (Weinheim an der Bergstrasse, Germany). 27 (2021) 10461. https://doi.org/10.1002/chem.202101242

[9] L. Pasquato, P. Pengo, P. Scrimin, Nanozymes: Functional nanoparticle-based catalysts, Supramolecular Chemistry. 17 (2005) 163-171. https://doi.org/10.1080/10610270412331328817

[10] A. Payal, S. Krishnamoorthy, A. Elumalai, J. Moses, C. Anandharamakrishnan, A Review on Recent Developments and Applications of Nanozymes in Food Safety and Quality Analysis, Food Analytical Methods. (2021) 1-22. https://doi.org/10.1007/s12161-021-01983-9

Materials Research Forum LLC
https://doi.org/10.21741/9781644901977-3

[11] J. Li, W. Liu, X. Wu, X. Gao, Mechanism of pH-switchable peroxidase and catalase-like activities of gold, silver, platinum and palladium, Biomaterials. 48 (2015) 37-44. https://doi.org/10.1016/j.biomaterials.2015.01.012

[12] I. Celardo, J.Z. Pedersen, E. Traversa, L. Ghibelli, Pharmacological potential of cerium oxide nanoparticles, Nanoscale. 3 (2011) 1411-1420. https://doi.org/10.1039/c0nr00875c

[13] M. Shamsipur, A. Safavi, Z. Mohammadpour, Indirect colorimetric detection of glutathione based on its radical restoration ability using carbon nanodots as nanozymes, Sensors and Actuators B: Chemical. 199 (2014) 463-469. https://doi.org/10.1016/j.snb.2014.04.006

[14] Z. Jin, N. Hildebrandt, Semiconductor quantum dots for in vitro diagnostics and cellular imaging, Trends in biotechnology. 30 (2012) 394-403. https://doi.org/10.1016/j.tibtech.2012.04.005

[15] F.P. Carvalho, Pesticides, environment, and food safety, Food and energy security. 6 (2017) 48-60. https://doi.org/10.1002/fes3.108

[16] J. Wei, L. Yang, M. Luo, Y. Wang, P. Li, Nanozyme-assisted technique for dual mode detection of organophosphorus pesticide, Ecotoxicology and environmental safety. 179 (2019) 17-23. https://doi.org/10.1016/j.ecoenv.2019.04.041

[17] H. Jia, D. Yang, X. Han, J. Cai, H. Liu, W. He, Peroxidase-like activity of the Co 3 O 4 nanoparticles used for biodetection and evaluation of antioxidant behavior, Nanoscale. 8 (2016) 5938-5945. https://doi.org/10.1039/C6NR00860G

[18] H. Jin, C. Guo, X. Liu, J. Liu, A. Vasileff, Y. Jiao, Y. Zheng, S.-Z. Qiao, Emerging two-dimensional nanomaterials for electrocatalysis, Chemical reviews. 118 (2018) 6337-6408. https://doi.org/10.1021/acs.chemrev.7b00689

[19] Y. He, W. Zhou, G. Qian, B. Chen, Methane storage in metal-organic frameworks, Chemical Society Reviews. 43 (2014) 5657-5678. https://doi.org/10.1039/C4CS00032C

[20] N.R. Nirala, R. Prakash, Quick colorimetric determination of choline in milk and serum based on the use of MoS 2 nanosheets as a highly active enzyme mimetic, Microchimica Acta. 185 (2018) 1-8. https://doi.org/10.1007/s00604-018-2753-2

[21] L. Huang, D.W. Sun, H. Pu, Q. Wei, Development of nanozymes for food quality and safety detection: Principles and recent applications, Comprehensive reviews in food science and food safety. 18 (2019) 1496-1513. https://doi.org/10.1111/1541-4337.12485

[22] G. Zhang, C. Zhu, Y. Huang, J. Yan, A. Chen, A lateral flow strip based aptasensor for detection of ochratoxin A in corn samples, Molecules. 23 (2018) 291. https://doi.org/10.3390/molecules23020291

[23] M. Hu, K. Korschelt, M. Viel, N. Wiesmann, M. Kappl, J.r. Brieger, K. Landfester, H. Therien-Aubin, W. Tremel, Nanozymes in nanofibrous mats with haloperoxidase-like activity to combat biofouling, ACS applied Materials & Interfaces. 10 (2018) 44722-44730. https://doi.org/10.1021/acsami.8b16307

[24] J. Huang, X.-L. Zhu, Y.-M. Wang, J.-H. Ge, J.-W. Liu, J.-H. Jiang, A multiplex paper-based nanobiocatalytic system for simultaneous determination of glucose and uric acid in whole blood, Analyst. 143 (2018) 4422-4428. https://doi.org/10.1039/C8AN00866C

[25] N. Stasyuk, G. Gayda, A. Zakalskiy, O. Zakalska, R. Serkiz, M. Gonchar, Amperometric biosensors based on oxidases and PtRu nanoparticles as artificial peroxidase, Food chemistry. 285 (2019) 213-220. https://doi.org/10.1016/j.foodchem.2019.01.117

[26] S. Wang, W. Deng, L. Yang, Y. Tan, Q. Xie, S. Yao, Copper-based metal-organic framework nanoparticles with peroxidase-like activity for sensitive colorimetric detection of Staphylococcus aureus, ACS applied materials & interfaces. 9 (2017) 24440-24445. https://doi.org/10.1021/acsami.7b07307

[27] Y. Li, X. You, X. Shi, Enhanced chemiluminescence determination of hydrogen peroxide in milk sample using metal-organic framework Fe-MIL-88NH 2 as peroxidase mimetic, Food Analytical Methods. 10 (2017) 626-633. https://doi.org/10.1007/s12161-016-0617-0

[28] A.H. Valekar, B.S. Batule, M.I. Kim, K.-H. Cho, D.-Y. Hong, U.-H. Lee, J.-S. Chang, H.G. Park, Y.K. Hwang, Novel amine-functionalized iron trimesates with enhanced peroxidase-like activity and their applications for the fluorescent assay of choline and acetylcholine, Biosensors and Bioelectronics. 100 (2018) 161-168. https://doi.org/10.1016/j.bios.2017.08.056

[29] T. Chen, X. Wu, J. Wang, G. Yang, WSe 2 few layers with enzyme mimic activity for high-sensitive and high-selective visual detection of glucose, Nanoscale. 9 (2017) 11806-11813. https://doi.org/10.1039/C7NR03179C

[30] H. Liu, Y.-N. Ding, B. Yang, Z. Liu, X. Zhang, Q. Liu, Iron doped CuSn (OH) 6 microspheres as a peroxidase-mimicking artificial enzyme for H2O2 colorimetric detection, ACS Sustainable Chemistry & Engineering. 6 (2018) 14383-14393. https://doi.org/10.1021/acssuschemeng.8b03082

[31] J. Rane, P. Adhikar, R. Bakal, Molecular imprinting: an emerging technology, Asian J Pharm Technol Innov. 3 (2015) 75-91.

[32] Z. Zhang, X. Zhang, B. Liu, J. Liu, Molecular imprinting on inorganic nanozymes for hundred-fold enzyme specificity, Journal of the American Chemical Society. 139 (2017) 5412-5419. https://doi.org/10.1021/jacs.7b00601

[33] F. Lin, T. Yushen, L. Doudou, W. Haoan, C. Yan, G. Ning, Z. Yu, Catalytic gold-platinum alloy nanoparticles and a novel glucose oxidase mimic with enhanced activity and selectivity constructed by molecular imprinting, Analytical Methods. 11 (2019) 4586-4592. https://doi.org/10.1039/C9AY01308C

[34] W. Guo, F. Pi, H. Zhang, J. Sun, Y. Zhang, X. Sun, A novel molecularly imprinted electrochemical sensor modified with carbon dots, chitosan, gold nanoparticles for the determination of patulin, Biosensors and Bioelectronics. 98 (2017) 299-304. https://doi.org/10.1016/j.bios.2017.06.036

[35] L. Qu, Y. Liu, J.-B. Baek, L. Dai, Nitrogen-doped graphene as efficient metal-free electrocatalyst for oxygen reduction in fuel cells, ACS nano. 4 (2010) 1321-1326. https://doi.org/10.1021/nn901850u

[36] R. Li, M. Zhen, M. Guan, D. Chen, G. Zhang, J. Ge, P. Gong, C. Wang, C. Shu, A novel glucose colorimetric sensor based on intrinsic peroxidase-like activity of C60-carboxyfullerenes, Biosensors and Bioelectronics. 47 (2013) 502-507. https://doi.org/10.1016/j.bios.2013.03.057

[37] C. Xu, X. Sun, X. Zhang, L. Ke, S. Chua, Photoluminescent properties of copper-doped zinc oxide nanowires, Nanotechnology. 15 (2004) 856. https://doi.org/10.1088/0957-4484/15/7/026

[38] S. Yao, C. Zhao, Y. Liu, H. Nie, G. Xi, X. Cao, Z. Li, B. Pang, J. Li, J. Wang, Colorimetric Immunoassay for the Detection of Staphylococcus aureus by Using Magnetic Carbon Dots and Sliver Nanoclusters as o-Phenylenediamine-Oxidase Mimetics, Food Analytical Methods. 13 (2020) 833-838. https://doi.org/10.1007/s12161-019-01683-5

[39] N. Fakhri, F. Salehnia, S.M. Beigi, S. Aghabalazadeh, M. Hosseini, M.R. Ganjali, Enhanced peroxidase-like activity of platinum nanoparticles decorated on nickel-and nitrogen-doped graphene nanotubes: colorimetric detection of glucose, Microchimica Acta. 186 (2019) 1-9. https://doi.org/10.1007/s00604-018-3127-5

[40] M. Jaishankar, T. Tseten, N. Anbalagan, B.B. Mathew, K.N. Beeregowda, Toxicity, mechanism and health effects of some heavy metals, Interdisciplinary toxicology. 7 (2014) 60. https://doi.org/10.2478/intox-2014-0009

[41] J. Li, Q. Cheng, H. Huang, M. Li, S. Yan, Y. Li, Z. Chang, Sensitive chemical sensor array based on nanozymes for discrimination of metal ions and teas, Luminescence. 35 (2020) 321-327. https://doi.org/10.1002/bio.3730

[42] Y. Zhou, B. Liu, R. Yang, J. Liu, Filling in the gaps between nanozymes and enzymes: challenges and opportunities, Bioconjugate chemistry. 28 (2017) 2903-2909. https://doi.org/10.1021/acs.bioconjchem.7b00673

[43] E. Golub, H.B. Albada, W.-C. Liao, Y. Biniuri, I. Willner, Nucleoapzymes: hemin/G-quadruplex DNAzyme-aptamer binding site conjugates with superior enzyme-like catalytic functions, Journal of the American Chemical Society. 138 (2016) 164-172. https://doi.org/10.1021/jacs.5b09457

[44] G.Y. Tonga, Y. Jeong, B. Duncan, T. Mizuhara, R. Mout, R. Das, S.T. Kim, Y.-C. Yeh, B. Yan, S. Hou, Supramolecular regulation of bioorthogonal catalysis in cells using nanoparticle-embedded transition metal catalysts, Nature chemistry. 7 (2015) 597-603. https://doi.org/10.1038/nchem.2284

[45] R. Antiochia, I. Lavagnini, F. Magno, Amperometric mediated carbon nanotube paste biosensor for fructose determination, Analytical letters. 37 (2004) 1657-1669. https://doi.org/10.1081/AL-120037594

[46] Z. Zhao, X. Chen, B. Tay, J. Chen, Z. Han, K.A. Khor, A novel amperometric biosensor based on ZnO: Co nanoclusters for biosensing glucose, Biosensors and Bioelectronics. 23 (2007) 135-139. https://doi.org/10.1016/j.bios.2007.03.014

[47] Y. Cheng, Y. Liu, J. Huang, Y. Xian, W. Zhang, Z. Zhang, L. Jin, Rapid amperometric detection of coliforms based on MWNTs/Nafion composite film modified glass carbon electrode, Talanta. 75 (2008) 167-171. https://doi.org/10.1016/j.talanta.2008.01.044

[48] C. Ozdemir, F. Yeni, D. Odaci, S. Timur, Electrochemical glucose biosensing by pyranose oxidase immobilized in gold nanoparticle-polyaniline/AgCl/gelatin nanocomposite matrix, Food Chemistry. 119 (2010) 380-385. https://doi.org/10.1016/j.foodchem.2009.05.087

[49] S. chuan Li, J. hua Chen, H. Cao, D. sheng Yao, Amperometric biosensor for aflatoxin B1 based on aflatoxin-oxidase immobilized on multiwalled carbon

nanotubes, Food Control. 22 (2011) 43-49.
https://doi.org/10.1016/j.foodcont.2010.05.005

[50] O.R. Miranda, X. Li, L. Garcia-Gonzalez, Z.-J. Zhu, B. Yan, U.H. Bunz, V.M. Rotello, Colorimetric bacteria sensing using a supramolecular enzyme-nanoparticle biosensor, Journal of the American Chemical Society. 133 (2011) 9650-9653. https://doi.org/10.1021/ja2021729

[51] R. Devi, S. Yadav, C. Pundir, Amperometric determination of xanthine in fish meat by zinc oxide nanoparticle/chitosan/multiwalled carbon nanotube/polyaniline composite film bound xanthine oxidase, Analyst. 137 (2012) 754-759. https://doi.org/10.1039/C1AN15838D

[52] S. Pal, M.K. Sharma, B. Danielsson, M. Willander, R. Chatterjee, S. Bhand, A miniaturized nanobiosensor for choline analysis, Biosensors and Bioelectronics. 54 (2014) 558-564. https://doi.org/10.1016/j.bios.2013.11.057

[53] H. Mahmoudi-Moghaddam, S. Tajik, H. Beitollahi, Highly sensitive electrochemical sensor based on La3+-doped Co3O4 nanocubes for determination of sudan I content in food samples, Food chemistry. 286 (2019) 191-196. https://doi.org/10.1016/j.foodchem.2019.01.143

[54] W. Wang, S. Gunasekaran, Nanozymes-based biosensors for food quality and safety, TrAC Trends in Analytical Chemistry. 126 (2020) 115841. https://doi.org/10.1016/j.trac.2020.115841

[55] S. Palanisamy, T. Kokulnathan, S.-M. Chen, V. Velusamy, S.K. Ramaraj, Voltammetric determination of Sudan I in food samples based on platinum nanoparticles decorated on graphene-β-cyclodextrin modified electrode, Journal of Electroanalytical Chemistry. 794 (2017) 64-70. https://doi.org/10.1016/j.jelechem.2017.03.041

[56] M.S.-P. Lopez, E. Redondo-Gómez, B. López-Ruiz, Electrochemical enzyme biosensors based on calcium phosphate materials for tyramine detection in food samples, Talanta. 175 (2017) 209-216. https://doi.org/10.1016/j.talanta.2017.07.033

[57] J. Wang, H. Wang, J. He, L. Li, M. Shen, X. Tan, H. Min, L. Zheng, Wireless sensor network for real-time perishable food supply chain management, Computers and Electronics in Agriculture. 110 (2015) 196-207. https://doi.org/10.1016/j.compag.2014.11.009

[58] L. Ma, Y. He, Y. Wang, Y. Wang, R. Li, Z. Huang, Y. Jiang, J. Gao, Nanocomposites of Pt nanoparticles anchored on UiO66-NH2 as carriers to construct

acetylcholinesterase biosensors for organophosphorus pesticide detection, Electrochimica Acta. 318 (2019) 525-533. https://doi.org/10.1016/j.electacta.2019.06.110

[59] M. Revenga-Parra, A.M. Villa-Manso, M. Briones, E. Mateo-Martí, E. Martínez-Periñán, E. Lorenzo, F. Pariente, Bioelectrocatalytic platforms based on chemically modified nanodiamonds by diazonium salt chemistry, Electrochimica Acta. 357 (2020) 136876. https://doi.org/10.1016/j.electacta.2020.136876

[60] F.T. T Cavalcante, I. R de A Falcão, J.E. da S Souza, T. G Rocha, I. G de Sousa, A.L. G Cavalcante, A.L. de Oliveira, M.C. M de Sousa, J. dos Santos, Designing of nanomaterials-based enzymatic biosensors: Synthesis, properties, and applications, Electrochem. 2 (2021) 149-184. https://doi.org/10.3390/electrochem2010012

[61] L. Huang, K. Chen, W. Zhang, W. Zhu, X. Liu, J. Wang, R. Wang, N. Hu, Y. Suo, J. Wang, ssDNA-tailorable oxidase-mimicking activity of spinel $MnCo_2O_4$ for sensitive biomolecular detection in food sample, Sensors and Actuators B: Chemical. 269 (2018) 79-87. https://doi.org/10.1016/j.snb.2018.04.150

[62] A.g. Molinero-Fernández, M. Moreno-Guzmán, M.A.n. López, A. Escarpa, Biosensing strategy for simultaneous and accurate quantitative analysis of mycotoxins in food samples using unmodified graphene micromotors, Analytical chemistry. 89 (2017) 10850-10857. https://doi.org/10.1021/acs.analchem.7b02440

[63] W.-J. Shen, Y. Zhuo, Y.-Q. Chai, R. Yuan, Cu-based metal-organic frameworks as a catalyst to construct a ratiometric electrochemical aptasensor for sensitive lipopolysaccharide detection, Analytical chemistry. 87 (2015) 11345-11352. https://doi.org/10.1021/acs.analchem.5b02694

[64] H. Yang, J. Zha, P. Zhang, Y. Qin, T. Chen, F. Ye, Fabrication of $CeVO_4$ as nanozyme for facile colorimetric discrimination of hydroquinone from resorcinol and catechol, Sensors and Actuators B: Chemical. 247 (2017) 469-478. https://doi.org/10.1016/j.snb.2017.03.042

[65] P. Weerathunge, R. Ramanathan, V.A. Torok, K. Hodgson, Y. Xu, R. Goodacre, B.K. Behera, V. Bansal, Ultrasensitive colorimetric detection of murine norovirus using NanoZyme aptasensor, Analytical chemistry. 91 (2019) 3270-3276. https://doi.org/10.1021/acs.analchem.8b03300

[66] C. Han, Q. Li, H. Ji, W. Xing, L. Zhang, L. Zhang, Aptamers: The Powerful Molecular Tools for Virus Detection, Chemistry-An Asian Journal. 16 (2021) 1298-1306. https://doi.org/10.1002/asia.202100242

Materials Research Forum LLC
https://doi.org/10.21741/9781644901977-4

Chapter 4

Nanomaterials Supported Enzymes: Environmental Applications for Depollution of Aquatic Environments

Fareeha Maqbool[1], Saima Muzammil[2], Muhammad Waseem[2], Tanvir Shahzad[3], Sabir Hussain[3], Muhammad Imran[4], Muhammad Afzal[1], Muhammad Hussnain Siddique*[1]

[1]Department of Bioinformatics and Biotechnology, Government College University, Faisalabad, Pakistan

[2]Department of Microbiology, Government College University, Faisalabad, Pakistan.

[3]Department of Environmental Sciences and Engineering, Government College University, Faisalabad, Pakistan

[4]Department of Environmental Sciences, COMSATS Institute of Information Technology, Vehari Campus, Pakistan

* mhs1049@gmail.com

Abstract

Increased pollution of worldwide water sources as well as difficulties in detecting and treating a wide range of contaminants impose significant health risks. Enzymes with their high activity and selectivity for chemical substrates are one of the promising options among the several technologies for the purification and depollution of aquatic environment. The operational performance of the enzymes is optimized through the immobilization process. Because of the unique physio-chemical properties of nanoparticles, they have become novel and attractive matrices for enzyme immobilization. Variety of composites consist of nanomaterials and enzymes have been discovered in order to improve enzyme stability, activity and functionality making nanosupported enzymes easier to use in depollution of aquatic environment. This chapter reveal different immobilization methods, nanosupports for immobilization and their uses in the depollution of aquatic environments.

Keywords

Enzyme, Immobilization, Nanoparticles, Biosensors, Covalent Attachment

Contents

1.   Introduction

The overall condition of the environment is intrinsically tied to the quality of life on the planet. Unluckily, advances in sciences and technologies results in production of vast amount of trash including from raw sewage to radioactive waste when released or thrown into the natural environment pose a serious threat to mankind's survival on earth. Water as a necessary component of life has influenced human history and culture. Abound 2.1 billion population does not have safe and pure drinking water, despite the development of advanced methods for the purification of water during the last few years and half of the world's population will live in such areas where water is scare by 2025 [1]. Water treatment technologies have the potential to supply purifying portable water to millions of rural people in a continuous, feasible, cost-effective, and flexible manner that have low environmental impact. Emerging pollutants (e.g., disruptors of endocrine, drugs used against pests and medicines), heavy metals and viruses are posing new challenges to present treatment technologies [2,3].

Enzymes are highly specific macromolecules that are used in various industries such as from bio-sensing to pharmaceutical and agrochemical synthesis, from industrial catalysis to biofuel and biofuel cell production and in the depollution and decontamination of aquatic environment. Because of their great specificity and selectivity enzyme-based procedures can produce higher yields and produce less hazardous byproducts than conventional chemical methods. Furthermore, enzymes can work under harsh conditions of temperature, pressure and pH as compared to traditional catalysts and resulting in significant energy and manufacturing cost reductions. But a lot of problems are associated with the development of enzyme-based technologies. One of the main problems is the cost of isolation and purification of enzymes and majority of the isolated enzymes work best in water-based conditions.

To address the constraints of enzyme-based applications and to assure high enzyme activity retention and functional stability enzyme immobilization is used. The chemical and physical properties of enzyme, as well as the support surfaces are used to determine which immobilization strategy to use. Enzymes have been entrapped in polymer matrices to achieve immobilization [4].

With the advancement of nanotechnology in recent years, numerous nanoparticles are produced for use in a variety of fields. Nanoparticles have a significant specific surface area but due to their small size they also have extraordinal optical, electrical, electronic, thermal, chemical, and mechanical properties as well as catalytic (ability to aid electron transfer) characteristics. When used for various chemical reactions, enzymes are immobilized on the nanoparticles to form enzyme nanoparticles (EnNPs) composites

Materials Research Foundations **126** (2022) 117-141   https://doi.org/10.21741/9781644901977-4

which have the enhanced enzyme stability, advantages of increased loading capacity of enzyme as well as the ease with which enzymes can be separated [5].

2. Enzymes

Enzymes are biomolecules that catalyze the particular chemical reactions. The specificity of enzymes depends on the 3-D shape of the enzymes. Enzymes temporarily bond to the substrate in a catalysis reaction and the enzyme–substrate complex is formed. Weak forces like Hydrogen bonding, van der Waals forces and hydrophobic interactions are necessary to form a combination of enzyme and substrate. The combination of enzyme and substrate determines a substrate's ability to be converted into a product. Enzymes are larger than the substrate. Enzymes have specific site where substrate can bind, and catalytic activity occur known as the active site. The final product is removed from the active site of the enzyme at the conclusion of the reaction, and the enzyme reverts to its original condition [6].

3. Sources of enzymes and their applications

Animals, plants and microorganisms are the main sources of enzymes. Enzymes like chymosin, pepsin, trypsin, etc. are obtained from animals, ficin, papain, bromelain, etc are obtained from plants furthermore microorganisms also produce pectinase, protease, cellulase, α-amylase or many other type of enzymes. Enzymes are also produced by microorganisms on the land and in the sea. From the industrial or biotechnological point of view, microbial enzymes are more preferable because these enzymes are obtained within short period of time with high catalytic activity [7].

Microorganisms are considered as ideal sources of enzyme because of fast reproduction rate, easy availability and range to produce many enzymes under desired conditions, manipulation and culturing in large quantities, cheap nutrients source and short fermentation times etc. [8–11]. The use of enzyme in different industries is increasing day by day due to the cost effectiveness, eco-friendly nature, low energy demand, non-toxicity, high catalytic activity, short processing time, high efficiency etc. [11–13].

The use of isolated enzymes in diverse field provided that enzymes are stabilized under high salty conditions, acidic pH, extreme temperature, surfactants along with alkalis. The majority of enzyme applications take place at high temperatures (For example, gelatinization of starch occur at 100 degrees Celsius, washing at 60–70 degrees Celsius and de-sizing of textiles at 80–90 degrees Celsius), under high quantity of salt, acidic conditions and particularly when surfactants are present [14].

Enzymes have a wide range of uses in aquatic environment like degradation of nitriles containing wastes, breaking of crude oil hydrocarbons as well as waste nitriles, water purification, disinfection of waste water and in the fields of industry like brewing, dairy, feed processing and beer industry, animal feed, leather and textile industry, pharmaceutical, paper and pulp industry. Drugs (41%) are the most common industries that use industrial enzymes, followed by feed and food (17%), detergents (17%), paper and leather (17%), and textiles (8 percent). Table 1 shows the global enzyme market's distribution across various application sectors.

Table 1 Advantages of enzymes in various industrial sectors.

Application field(s)	Enzymes	Advantages	References
Pulp and paper industry	Amylase	Improvement in drainage	[11,12]
	Laccase	Improved brightness through bleaching	
	ß-xylanase	Bleach boosting	
Waste management	Amidase	Degradation of nitriles containing wastes	[11]
	Lipase	Breaking of crude oil hydrocarbons	
	Nitrile hydratase	Dedragadation of nitriles containing wastes	
Textile industry	Cellulase	Cotton softening, denim finishing	[12,87]
	Pectinase	Alkaline bio-scouring of cellulosic textiles	
	Laccase	Dyeing of fabrics and Chlorine-free bleaching	[11,12]
Animal-feeds industry	Xylanase	used as degradable fiber in viscous liquids	[12,87]
	Phytase	Cause the breakdown of phytic acid to obtain calcium, phosphorus and Mg ions	
	α- amylase	Hydrolyzes starch	
Food processing industry	Cellulase Hemicellulase	Fruit liquefaction	[12,87]
	Dextranase	Hydrolyze the dextran(a byproduct of sugar production) increases the flow's viscosity and lowers industrial recovery	

4.   Enzyme immobilization

Enzymes are the naturally occurring highly activated biomolecules that speed up the chemical reactions. They have capacity for use in different fields due to many advantages such as ability to catalyze a reaction at diverse temperature and pH, easy and fast production rate, high efficiency, environmentally sustainable, non-toxicity, biodegradability and biocompatibility [15–17]. All these characteristics make the enzymes popular substitute to traditional chemical catalysts in the conditions that are reasonable and eco-friendly [18,19].

In their native context enzymes bind to cell membranes and remain active and stable there. Structure immobilization is comparable to enzyme immobilization. As a result, enzymes attached to a solid surface are highly stable and resistant to environmental changes than the free enzymes. More importantly, the immobilization of enzymes lead towards the formation of heterogenous system that allows the reuse of enzymes, indefinitely production rate, simple reaction termination, easy recovery of enzyme and various reactor designs. Immobilization also prevents the products from blocking the enzyme, resulting in improved functional characteristics [20].

Enzyme immobilization to a solid support not only make it considerably easier to extract enzyme from the reaction media but also retain the enzyme activity. Once enzymes are immobilized, the flexibility of their tertiary structure is locked by binding with a solid support as a result their stability is enhanced. Immobilization process involves the satisfaction of following circumstances: **a)** supported material used for enzyme immobilization should be used again and again **b)** biocompatible **c)** have a large surface area per unit volume [21] for a suitable enzyme loading. Stability and loading capacity are the most important characteristics of an immobilized enzyme to be used in biotechnology and industrial field [22]. Immobilized Enzymes can prevent the degradation, accumulation, denaturation at different Heat and pH values. Moreover, biocompatibility of immobilization method determine the biocatalytic activity of enzyme and allow the enzyme to recognize substrates and cofactors present in the reaction [23].

From the last few years various enzyme immobilization methods have been developed that are more efficient and can be used in a variety of industry such as in the food industry, drug metabolism, biodiesel and bioethanol synthesis, biosensor production, antibiotic production and bioremediation. Immobilized enzymes are widely used because they are environmentally safe, cost-effective and considerably easier to utilize [24].

Materials Research Foundations **126** (2022) 117-141   https://doi.org/10.21741/9781644901977-4

5. Methods of Immobilization

In 1916, Nelson and Griffin reported first immobilized enzyme that was immobilized through simple adsorption on an artificially synthesized aluminum hydroxide carrier [25,26]. From 1960 enzyme immobilization on different carriers have been started through covalent attachment and entrapment [25,27]. Then in the late 1960s the use of polysaccharide beads having pores such as dextran, agarose or cellulose have been started for enzyme immobilization to improve enzyme stability and quantity of enzyme loaded onto a carrier [27]. Following methods are used for the immobilization of enzymes 1) Adsorption 2) Entrapment 3) Covalent binding 4) Cross-linking.

5.1 Adsorption

This is simple and easy technique used for immobilization. Adsorption methods include ion adsorption, physical adsorption and affinity adsorption. Physical adsorption depend upon the hydrogen bonding, van der Waals forces, electrostatic or hydrophobic interactions for the attachment of enzyme on insoluble solid support [28,29], or the immobilization of enzyme in mesoporous materials' pores [30,31]. The main advantage of Physical adsorption is to prevent the deactivation of enzymes usually induced by the chemical modifications to the enzyme surfaces such as in the cross-linking and covalent linkage approaches. The insoluble solid supports used for the enzyme immobilization include ceramics, silica gel, anion exchange resins, alumina, cation exchange resins, controlled pore glass. Physical adsorption of lipase enzyme on the Phenyl-Sepharose CL-4B and Octyl-Sepharose CL-4B resins by hydrophobic adsorption was reported by Bastida and colleagues in 1998. Another example of this technique is the adsorption of enzyme laccase into the pores and the surfaces of micromesoporous Zr-metal organic frameworks (Zr-MOFs) [30]. This method is cost effective thus no additional modifications and binding chemicals are required. There are following disadvantages of adsorption method **1)** low stability of immobilized enzyme due to the weak and reversible attachment between the enzyme and solid-support [24] **2)** easily desoption as a result of fluctuation in temperature and variations in the concentrations of ions and substrate [32].

5.2 Entrapment

Entrapment, is an encapsulation method in which enzymes are trapped inside inert and porous material. In entrapment, movement of enzyme is restricted within a gel having pores though maintaining the independence of molecules in the solution. Encapsulation in the fibers and gels is an easy and suitable approach because it involve the low molecular weight substrate and product [32]. In entrapment material used for the support of enzyme include agar, calcium alginate, chitosan, polyvinyl alcohol, cellulose triacetate, polyacrylamide, collagen, gelatin, polyurethane and silicone rubber [24]. Synthetic polymers such as polyvinyl alcohol hydrogel [33], polyacrylamide [34] have also been investigated.

5.3 Covalent binding

In this method, covalent bond is formed between the activated functional groups of enzymes and support materials through chemical reaction. Covalent association between enzymes and support matrix occurs through amino acids (Histidine, Aspartic acid and Arginine) present on the side chain while different functional groups such as phenolic, indolyl, imidazole, hydroxyl, and others influence the degree of reactivity [35]. Enzyme infiltration from the support matrix can be prevented via covalent bond. So, enzymes remain fixed [36]. Enzymes get more rigid structure through covalent attachment [37].

Durability of the enzyme depend on the excess number of bounds and the length of the bound between the support surface and enzyme[24]. Sometimes the activity of the functional groups on the support matrix need to be activated by using different chemicals after which enzymes are able to couple with the support surfaces through covalent bonding. Cyanogen bromide (CNBr)- agarose and CNBr-activated-Sepharose containing carbohydrate moiety and glutaraldehyde as a spacer arm have imparted thermal stability to covalently bound enzymes [38,39].

5.4 Cross-linking

Immobilization through cross-linked method was first reported by Richards and Quiocho for cross-linking carboxypeptidase [40] and by Habeeb for cross-linking trypsin [41]. In cross-linked method enzyme molecules are attached to each other through via cross-linking reagent e.g., the most popular of which is glutaraldehyde. Often a small amount of an inactive protein such as Bovine serum albumin also used as a cross-linking reagent in case of susceptible enzymes [26,42]. In cross-linked method enzymes are not attached to solid support, so they are usually firm. Since this method involves covalent kind of bound so biocatalyst immobilized in this method usually go through conformational changes that cause the loss of enzyme activity. The main disadvantages of this method are:

1) Undesirable loss of activity which is due to the involvement of catalytic groups in the interaction is responsible for immobilization.

2) This technique is challenging to regulate, because bigger enzyme aggregates that shows higher activity are laborious to gain.

3) The preparation of immobilized enzyme is not so easy because of the gelatinous nature which make less use of the cross-linked method.

4) The toxicity of the reagents used for cross-linking is a barrier to many enzymes using this approach [32].

Enzyme nanoparticle composites have been created using this technology such as horseradish peroxidase (HRP) [43], trypsin [44] , uricase [45], penicillin acylase [42,46], cholesterol oxidase (ChOx) [47] and lipase [42,48] with the benefits of forming a stable and re-usable biocatalyst at a lower production cost.

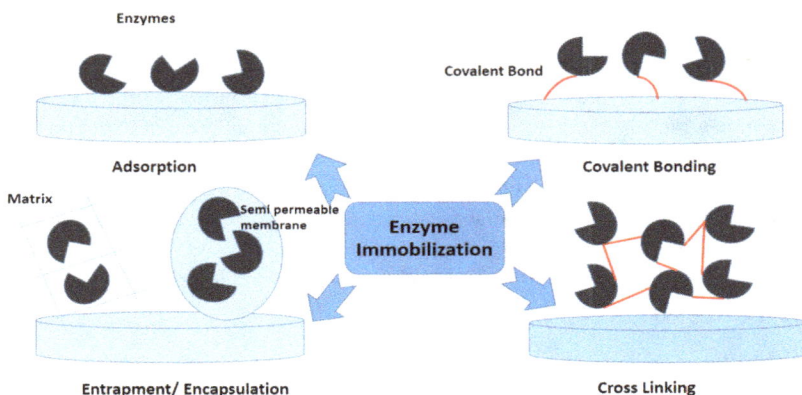

Fig 1. Enzyme immobilization methods.

6.   Nanosupports for enzyme immobilization

6.1   Silica nanosupports

The most widely abundant thermally and chemically inert solid support used for enzyme immobilization is Silica. Surfactant-template polymerization processes convert the silicon alkoxide into the mesoporous silica materials. This procedure is carried out under severe circumstances including organic solvents, high temperature, and extreme pH that cause the loss of enzyme activity. These mesoporous silica particles gain a great importance because these have enormous surface areas (300−1200 m2 /g) and holes (1.6 to 30 nm) to attach size varying   polypeptides [49–51]. Covalent bond binds the enzyme to the bare mesoporous silica or chemically grafted with the functional groups like Thiol, Amines and Silanols. Enzymes can also be captured by the silica particles when the silica is synthesized in the presence of enzymes. At neutral pH of water, enzymes can also be entrapped by the silica particles via biosilicification (a process occur in the presence of biomolecules) [52].

6.2   Carbon nanosupports

There are two types of carbon nanosupports carbon nanotubes (CNTs) and graphene. Enzymes can bound to these supports through physical adsorption or by surface grafting [53]. Polymerization and carbonization convert the carbon precursor molecules into mesoporous carbon materials. In contrast with the silica nanosupports, carbon nanosupports gain a great importance in the field of enzyme-based amperomatric sensors because Carbon nanosupports are electrically conductive and can facilitate electron transfer processes on electrodes. [51,53]. Carbon nanosupports can be made through a method known as nanogel approach. In this method a permeable and thin polymer layer is produced which surround the enzyme molecules depend on a two-step process. In the 1st step by altering amine groups in amino acid residues, ethenyl groups are brought to the surface of enzyme. After adding monomers, cross-linkers and initiators process of polymerization occur at ethenyl groups that cause the trapping of single enzyme molecules in the polymer. The activity of enzymes is only slightly affected by nanogel encapsulation yet their resilience is maintained even under extreme conditions. As a result, it allows for further enzymes modification such as immobilization under severe conditions that would kill the enzymes present freely [54].

6.3   Metallic nanosupports

Metal-organic frameworks (MOFs) are permeable metal nanoparticles made through the cross-linking of organic ligands and metal ions. Enzymes are immobilized on the MOFs through covalent binding or by physical adsorption. Covalent bound is formed between the molecules of enzymes and built-in functional groups of MOFs [55]. MOFs arrange themselves around the molecules of enzyme in a one-pot manufacturing procedure [56]. Another type of metallic nanosupport is protein inorganic hybrid nanoflowers. The first was made by coprecipitating copper salts and proteins [57] in a single process, and later it was broaden to other metal ions such as iron [58].

The main advantage of hybrid nanoflowers and MOFs is that they can produce a microenvironment associated with high concentration of metal ion, that could interact collectively with the enzymes that are immobilized and increase the activity of enzymes, especially for enzymes containing metals [56–58]. Enzymes have been immobilized using magnetic nanoparticles enabling easy recovery and separation from the reaction media [53].

Enzymes have also been attempted to be immobilized on native metal nanoparticles. Biological synthesis, dendrimer-assisted synthesis and a variety of other techniques have been discovered to regulate capacity and shape of native-metal nanomaterials [59]. Native-metals are often minor employed in water treatment because to their extreme prices, but optical characteristics depend on the state of aggregation in combination with enzyme activity can be exploited to develop sensors for the pollutant detection [60]. As summarized in table 2 unique benefits of nanomaterials when used as an immobilization support.

Table 2 Nanomaterials and their benefits as an immobilized support.

Nanomaterials	Benefits	Applications	References
Silica synthesis using chemicals	Easily synthesis	Removal of pollutants, disinfection	[52,65,73]
Silica synthesis using biological system	Conditions for synthesis are milder Minimum loss of enzyme activity	Disinfection, removal of pollutants	
Grapheme, mesoporous carbon and CNTs	Unique electrical conductivity	Amperometric sensors based on enzymes	[74,77,81,88]
Nanogels	Minimal enzyme activity loss	Contaminant removal	[54]
Magnetic nanoparticles	Easy and simple recovery	Deportation of pollutants	[53]
Noble metal Nanoparticles	Extraordinary optical feature that is depending on the aggregation state	Colorimetric sensors based on enzymes	[60,86,89]

7.   Applications of nanosupported enzymes in the depollution of aquatic environment

7.1   Water treatment applications

7.1.1   Eradication of emerging pollutants

Newly emerging pollutants are present everywhere in the aquatic environment but sometimes these are not removed via the traditional methods of water treatment like filtration, sedimentation and coagulation. Many modern approaches including as photocatalysis and electrochemistry have been developed and employed to improve the effectiveness of these compounds' removal. The enzymatic approach for example has a comparatively high effectiveness, polyvalence and consume minimal chemical or energy [61–63].

So far several enzymes including peroxidases, laccase and organophosphates hydrolase a nd atrazine chlorohydrolase were discovered and developed to remove a number of pollutants such as: disruptors of endocrine, pesticides, drugs and cosmetics [64,65]. The main   problem of enzyme   assisted   methods   is that   mobilized   enzymes might rapidly deactivated due to the sensitivity of metal ions and ligands into the water as well as the outcomes of their own operations. But the problem could be solved through the immobilization of enzymes on nanocarriers that usually enhance the activity and functionality of enzymes with a minimum activity loss [66]. During a project, a membrane reactor with the laccase enzyme that is immobilized on silica nanoparticles withdrawn about 66 percent of power supply bisphenols A [65].

Metal nanocarriers have shown that they improve the activity of immobilized enzymes [57,58,67]. For example, 2,4-dichlorophenol hydroxylase immobilization in copper nanoflowers resulted in a 160 percent increase in 2,4- dichlorophenol degradation [67]. For enzymes with metal ion-containing cofactors, the improvement impact is more significant. Immobilized HRP with iron nanoflowers and copper nanoflowers were found to be about three-times more stable and active as compared to mobilized HRP in a study of the heme-containing enzyme horseradish peroxidase (HRP) [58].

Pollutants can be adsorbed and concentrated near the enzymes surfaces and in the microenvironment around them using nanosupports, improving apparent enzyme kinetics and facilitating contaminant clearance. For example, a supernegatively charged inner surface of a lumazine synthase protein cage was used to collect substrates containing +ive charge over the substrates containing –ive and neutral charges, inverting substrate selectivity of encased enzyme by 480 times [68]. Vault nanoparticles with lipophilic properties were created by genetically modifying the N-terminus of main vault protein by attaching an amphipathic peptide to that terminus [69].

The advantages of immobilizing enzymes on nanosupports also extend to the use of enzyme couples. Certain pollutants (such as $C_3H_5Cl_3$) need gradually conversion involving numerous enzymes [70], whose effectiveness become limited by the diffusion of substrates between the enzymes. Enzymes' collocation on nanosupports' tiny surfaces might decrease the internal diffusion path, improving all over kinematics. Copper nanoflowers with immobilized HRP and immobilized glucose oxidase had 5 times higher activity as compared to mobilized enzymes and a 60% greater feedback compared to the combination of copper nanoflowers with immobilized HRP and immobilized glucose oxidase [71].

7.1.2 Disinfection

Chlorination and Ozonation which are currently used to disinfect drinking water can yield hazardous byproducts. Antimicrobial enzymes on the other hand attack microorganisms and biofilms (produced by bacteria) through adhesion or by disrupting the components of bacterial cells limiting the creation of any toxic byproducts[72]. Due to the cleavage of crosslinking bonds present in peptidoglycans (a significant constituent of Gram +ve cell walls) lysostaphin and lysozyme are effective against Gram +ve bacteria [73,74]. Another option is the use of such enzymes that produce oxidative stresses such as hydrogen peroxide that inhibit the growth of bacteria [75]. Bacteria attach with one another and on the other surfaces using the adhesion proteins Proteases and the hydrolysis of these proteins reduces the microbial fouling [72]. Antimicrobial enzymes such as subtilisin, bacteriophage lysine and lysosomal extract are also examples[72,76]. Antimicrobial enzymes that have been immobilized improve their stability and prevent them from being digested by bacteria, making them easier to utilize in reactors that may be utilized as separate purification units and to control biofouling in various point-of-use (POU) procedures [73].

Furthermore, conjugating antimicrobial enzymes to nanosupports protects the activity of enzymes throughout subsequent processing thus expanding the range of applications for antimicrobial enzyme nanomaterial composites. Protease immobilized with carbon nanotubes, for example have been integrated into a matrix of polymer to inhibit microbial fouling that might utilize in sense of coatings on the interior sides of tanks used for water storage in order to enhance safety of water   [77].

7.2   Water monitoring applications

7.2.1   Electro-enzymatic method

Biosensors constructed on the basis of electro-enzymatic approach are commonly used and these sensors use the electrodes on which enzymes are coated to detect the pollutants. Voltage is produced when potential is applied to the species made or consumed by enzymatic activity and activity is determined by the pollutant concentration. But the loss of enzyme stability and activity are two issues that could restrict the efficacy of this type of sensor. A combination of enzymes and nanosupports having conductivity like metal nanoparticles (gold nanoparticles) furthermore carbon nanomaterials (e.g., CNTs, graphene) improves the constancy and carrying capacity of enzymes while promoting e⁻ transport on conductors resulting in susceptible and fast-responding detectors [78].

Two main detection methodologies have been established that depend on either contaminants interact directly with the enzyme or have an indirect effect on enzyme activity. For starters pollutants produced under the catalysis of enzymes are identified by using amperometry. Paraoxon was identified at 0.15 μM using an electrode that is coated with the carbon nanotubes conjugated or immobilized with the organophosphate hydrolase enzyme greater than 10-times improvement than the electrode coated with the free-enzyme [79]. With the suitable nanosupported enzymes other pollutants including phenol [80], bisphenol A [81], and fenitrothion [82] have been identify. This method has a high selectivity over interfering substances due to the high specificity of enzymes. As a result even in mixtures pollutants can be recognized at very low quantities. Second, the nonenzyme-reactive contaminants' inhibitory effects on enzyme activity can be utilized to identify the pollutants.

Many enzymes are inhibited by heavy metal ions due to the blockage of sulfhydryl groups of enzymes causing mercaptides to develop [83]. A conductor coated with the free enzyme is 2 time less sensitive to Hg^{2+} than the conductor covered with immobilized urease enzyme on gold nanoparticles, detecting $Hg2^+$ at minimum concentration such as 50 nM [84]. The inhibitory technique considerably increases pollutant detection capacity however it is not so much susceptible and selective to the complex components of natural water samples.

7.2.2 Colorimetric method

This approach has been used to determine contamination levels quickly. They are less sensitive than electro-enzymatic sensors but they are easier to use by the general population. The production of colored compounds as a result of contaminant conversion mediated by enzymes can be utilized to directly estimate concentrations. Immobilization of enzymes with nanomaterials enhance enzymes activity and stability resulting in speedy reactions and lower detection limit, as well as making it easier to transfer the methodology from the lab to the field applications [59,90]. Noble metal nanoparticles' distinctive surface and optical features are especially important in the development of colorimetric enzyme sensors. Nanometals can increase the quantum efficiency of fluorophores that are bound to the surface. A fluorophore (competitive enzyme inhibitor) could be attached to enzyme that is immobilized furthermore persist on the nanometal particle's exterior part resulting in increased radiance [71].

7.2.3 Bacterial monitoring

During the detection of bacteria when immobilized enzyme creates signal interactions of nanosupported enzymes with cell surfaces occur. When β D-galactopyranoside (chlorophenol red) undergo hydrolysis, this reaction is catalyzed by the β galactosidase that cause the change in color and it is known as a chromogenic reaction. The anionic galactosidase is adsorbed and inhibited by the modified gold nanoparticles with the ligands that are cationic in nature. Immobilized galactosidase is shifted when anionic bacterium surfaces bind to nanoparticles, restore its ability to catalyze the colorimetric reactions when CPRG is present. A strip test is constructed depend upon the galactosidase enzyme that is immobilized has been used to identify the bacteria at minimum concentrations such as 10^4 bacteria/mL [86]. But due to the interactions with all anionic surfaces this approach is unable to distinguish infections from nonpathogenic strains. The specificity should be improved by applying enzymes that attack specific extracellular chemicals or components of the cell of harmful microbes [71].

Conclusion and Future Perspectives

Nanosupported enzymes are promising technologies used for the depollution of aquatic environment. In both aqueous and non-aqueous conditions, the immobilization procedure has been employed to improve enzyme activity and stability. In enzyme immobilization the support matrix must be carefully chosen and designed. Because of their tiny size and huge surface area nanoparticles have recently emerged as a useful technique for developing effective supports for enzyme stabilization [32].

In addition to providing a flexible platform for tailoring enzymes and nanomaterials to provide desirable water treatment biodegradable immobilized enzymes can provide productive, eco-friendly and unpolluted methods for water purification. Enzymes used against for the microbes assisted by nanoparticles are a safe disinfection alternative that can help to increase the safety of water storage tank. High quality sensors constructed on the basis of nanosupported enzymes allow for on-the-spot pollution detection.

However issues such as adaptability, potential health hazards, and expenses are impeding the technology's transition from the lab to the field. However despite the promising results numerous unforeseen disadvantages of enzyme onto nanoparticles (EnNPs) including as aggregation, precipitation and bio-incompatibility remain a source of concern in terms of health, environment and the economy [5]. It is necessary to investigate many additional enzymes in order to broaden their use in pollutant removal.

The use of pure enzymes in immobilization on the other hand comes at a high cost. The vault nanoparticles for example have been found to trap enzymes through particular interactions. They can specifically capture certain enzymes removing the need for enzyme purification. The nanosupported enzyme systems will profit economically from the development of nanoparticles with multiple functions that cause the purification and immobilization the enzymes at the same time [71].

References

[1] Progress on Drinking Water, Sanitation and Hygiene: 2017 update and SDG baselines - UNICEF DATA, (2017). https://data.unicef.org/resources/progress-drinking-water-sanitation-hygiene-2017-update-sdg-baselines/ (accessed August 22, 2021).

[2] M.A. Montgomery, M. Elimelech, Water and sanitation in developing countries: Including health in the equation - Millions suffer from preventable illnesses and die every year, Environ. Sci. Technol. 41 (2007) 17-24. https://doi.org/10.1021/es072435t

[3] E. Mintz, J. Bartram, P. Lochery, M. Wegelin, Not Just a Drop in the Bucket: Expanding Access to Point-of-Use Water Treatment Systems, Am. J. Public Health. 91 (2001) 1565. https://doi.org/10.2105/AJPH.91.10.1565

[4] A. Steven Campbell, A. Steven, Graduate Theses, Dissertations, and Problem Reports 2013 A systematic study of enzyme-nanomaterial interactions for A systematic study of enzyme-nanomaterial interactions for application in active surface decontamination application in active surface deco, (2013). https://researchrepository.wvu.edu/etd/538 (accessed August 22, 2021).

[5] C.-K. Lee, A.-N. Au-Duong, Enzyme immobilization on nanoparticles: Recent Application, Emerg. Areas Bioeng. First Ed. (2018). https://doi.org/10.1002/9783527803293.ch4

[6] N. Arabacı, T. Karaytuğ, A. Demirbas, I. Ocsoy, A. Katı, Nanomaterials for Enzyme Immobilization, Green Synth. Nanomater. Bioenergy Appl. (2020) 165-190. https://doi.org/10.1002/9781119576785.ch7

[7] S. VAIDYA, P. Srivastava, P. RATHORE, A. Pandey, Amylases: a Prospective Enzyme in the Field of Biotechnology, J. Appl. Biosci. 41 (2015) 1-18. https://www.researchgate.net/publication/278670695 (accessed August 4, 2021).

[8] A. Illanes, A. Cauerhff, L. Wilson, G.R. Castro, Recent trends in biocatalysis engineering, Bioresour. Technol. 115 (2012) 48-57. https://doi.org/10.1016/j.biortech.2011.12.050

[9] B. Kalpana, S.P.-J. of basic microbiology, undefined 2014, Halotolerant, acid-alkali stable, chelator resistant and raw starch digesting α-amylase from a marine bacterium Bacillus subtilis S8-18, Wiley Online Libr. 54 (2014) 802-811. https://doi.org/10.1002/jobm.201200732

[10] L. Liu, H. Yang, H. Shin, R. Chen, J. Li, … G.D.-, undefined 2013, How to achieve high-level expression of microbial enzymes: strategies and perspectives, Taylor Fr. 4 (2013) 212-223. https://doi.org/10.4161/bioe.24761

[11] R. Singh, M. Kumar, A. Mittal, P.K. Mehta, Microbial enzymes: industrial progress in 21st century, 3 Biotech. 6 (2016). https://doi.org/10.1007/s13205-016-0485-8

[12] S. Li, X. Yang, S. Yang, M. Zhu, X. Wang, Technology prospecting on enzymes: Application, marketing and engineering, Comput. Struct. Biotechnol. J. 2 (2012) e201209017. https://doi.org/10.5936/csbj.201209017

[13] J.M. Choi, S.S. Han, H.S. Kim, Industrial applications of enzyme biocatalysis: Current status and future aspects, Biotechnol. Adv. 33 (2015) 1443-1454. https://doi.org/10.1016/j.biotechadv.2015.02.014

[14] P. V. Iyer, L. Ananthanarayan, Enzyme stability and stabilization-Aqueous and non-aqueous environment, Process Biochem. 43 (2008) 1019-1032. https://doi.org/10.1016/j.procbio.2008.06.004

[15] R. Pieroni Vaz, L. Rios De Souza, M.E. Ximenes, F. Filho, Title: AN OVERVIEW OF HOLOCELLULOSE-DEGRADING ENZYME IMMOBILIZATION FOR USE IN BIOETHANOL PRODUCTION, "Journal Mol. Catal. B, Enzym. (2016).

[16] N. SS, R. VK, Magnetic-metal organic framework (magnetic-MOF): A novel platform for enzyme immobilization and nanozyme applications, Int. J. Biol. Macromol. 120 (2018) 2293-2302. https://doi.org/10.1016/j.ijbiomac.2018.08.126

[17] J. Cui, S. Ren, B. Sun, S.J.-C.C. Reviews, U. 2018, Optimization protocols and improved strategies for metal-organic frameworks for immobilizing enzymes: Current development and future challenges, Elsevier. (2018). https://www.sciencedirect.com/science/article/pii/S0010854517307026 (accessed August 5, 2021).

[18] J. Cui, S. Jia, Organic-inorganic hybrid nanoflowers: A novel host platform for immobilizing biomolecules, Coord. Chem. Rev. 352 (2017) 249-263. https://doi.org/10.1016/j.ccr.2017.09.008

[19] W. Feng, P. Ji, Enzymes immobilized on carbon nanotubes, Biotechnol. Adv. 29 (2011) 889-895. https://doi.org/10.1016/j.biotechadv.2011.07.007

[20] B. Krajewska, Application of chitin- and chitosan-based materials for enzyme immobilizations: A review, Enzyme Microb. Technol. 35 (2004) 126-139. https://doi.org/10.1016/j.enzmictec.2003.12.013

[21] S.A. Ansari, Q. Husain, Potential applications of enzymes immobilized on/in nano materials: A review, Biotechnol. Adv. 30 (2012) 512-523. https://doi.org/10.1016/j.biotechadv.2011.09.005

[22] S. Cao, P. Xu, Y. Ma, X. Yao, Y. Yao, M. Zong, X. Li, W. Lou, Recent advances in immobilized enzymes on nanocarriers, Cuihua Xuebao/Chinese J. Catal. 37 (2016) 1814-1823. https://doi.org/10.1016/S1872-2067(16)62528-7

[23] M. Bilal, T. Rasheed, Y. Zhao, H.M.N. Iqbal, J. Cui, "Smart" chemistry and its application in peroxidase immobilization using different support materials, Int. J. Biol. Macromol. 119 (2018) 278-290. https://doi.org/10.1016/j.ijbiomac.2018.07.134

[24] N. Arabacı, T. Karaytuğ, A. Demirbas, I. Ocsoy, A. Katı, Nanomaterials for Enzyme Immobilization, Green Synth. Nanomater. Bioenergy Appl. (2020) 165-190. https://doi.org/10.1002/9781119576785.ch7

[25] L. Cao, L. van Langen, R.S.-C. opinion in Biotechnology, undefined 2003, Immobilised enzymes: carrier-bound or carrier-free?, Elsevier. (n.d.). https://www.sciencedirect.com/science/article/pii/S095816690300096X (accessed August 9, 2021).

[26] Y. Zhang, J. Ge, Z. Liu, Enhanced Activity of Immobilized or Chemically Modified Enzymes, ACS Catal. 5 (2015) 4503-4513. https://doi.org/10.1021/acscatal.5b00996

[27] B.M. Brena, F. Batista-Viera, Immobilization of Enzymes, (2006) 15-30. https://doi.org/10.1007/978-1-59745-053-9_2

[28] J.C. Quilles Junior, A.L. Ferrarezi, J.P. Borges, R.R. Brito, E. Gomes, R. da Silva, J.M. Guisán, M. Boscolo, Hydrophobic adsorption in ionic medium improves the catalytic properties of lipases applied in the triacylglycerol hydrolysis by synergism, Bioprocess Biosyst. Eng. 39 (2016) 1933-1943. https://doi.org/10.1007/s00449-016-1667-9

[29] A. Bastida, P. Sabuquillo, P. Armisen, R. Ferná Ndez-Lafuente, J. Huguet, J.M. Guisá, A Single Step Purification, Immobilization, and Hyperactivation of Lipases via Interfacial Adsorption on Strongly Hydrophobic Supports, Biotechnol Bioeng. 58 (1998) 486-493. https://doi.org/10.1002/(SICI)1097-0290(19980605)58:5<486::AID-BIT4>3.0.CO;2-9

[30] S. Pang, Y. Wu, X. Zhang, B. Li, J. Ouyang, M. Ding, Immobilization of laccase via adsorption onto bimodal mesoporous Zr-MOF, Process Biochem. 51 (2016) 229-239. https://doi.org/10.1016/j.procbio.2015.11.033

[31] J. Fan, J. Lie, L. Wang, C. Yu, B. Tu, D. Zhao, Rapid and high-capacity immobilization of enzymes based on mesoporous silicas with controlled morphologies, Chem. Commun. 3 (2003) 2140-2141. https://doi.org/10.1039/b304391f

[32] R. Ahmad, M. Sardar, Enzyme Immobilization: An Overview on Nanoparticles as Immobilization Matrix, (2015). https://doi.org/10.4172/2161-1009.1000178. https://doi.org/10.4172/2161-1009.1000178

[33] G. Z, R. M, R. M, S. M, S. B, Entrapment of beta-galactosidase in polyvinylalcohol hydrogel, Biotechnol. Lett. 30 (2008) 763-767. https://doi.org/10.1007/s10529-007-9606-0

[34] A. Deshpande, S.F. D'souza, G.B. Nadkarni, Coimmobilization of D-amino acid oxidase and catalase by entrapment ofTrigonopsis variabilis in radiation polymerised Polyacrylamide beads, J. Biosci. 1987 111. 11 (1987) 137-144. https://doi.org/10.1007/BF02704664

[35] S.D.-C. Science, U. 1999, Immobilized enzymes in bioprocess, JSTOR. (1999). https://www.jstor.org/stable/24102915 (accessed August 10, 2021).

[36] D.M. Liu, J. Chen, Y.P. Shi, Advances on methods and easy separated support materials for enzymes immobilization, TrAC - Trends Anal. Chem. 102 (2018) 332-342. https://doi.org/10.1016/j.trac.2018.03.011

[37] M.C.P. Gonçalves, T.G. Kieckbusch, R.F. Perna, J.T. Fujimoto, S.A.V. Morales, J.P. Romanelli, Trends on enzyme immobilization researches based on bibliometric analysis, Process Biochem. 76 (2019) 95-110. https://doi.org/10.1016/j.procbio.2018.09.016

[38] V. Singh, M. Sardar, M.G.-I. of E. and Cells, undefined 2013, Immobilization of enzymes by bioaffinity layering, Springer. (n.d.). https://link.springer.com/content/pdf/10.1007/978-1-62703-550-7_9.pdf (accessed August 10, 2021). https://doi.org/10.1007/978-1-62703-550-7_9

[39] H. M, K. X, Immobilization of enzymes on porous silicas--benefits and challenges, Chem. Soc. Rev. 42 (2013) 6277-6289. https://doi.org/10.1039/c3cs60021a

[40] F. Quiocho, … F.R. the N.A. of S. of, undefined 1964, Intermolecular cross linking of a protein in the crystalline state: carboxypeptidase-A, Ncbi.Nlm.Nih.Gov. (1964). https://www.ncbi.nlm.nih.gov/pmc/articles/PMC300354/ (accessed August 14, 2021).

[41] A.F.S.A. Habeeb, Preparation of enzymically active, water-insoluble derivatives of trypsin, Elsevier. (1967) 264-268. https://doi.org/10.1016/0003-9861(67)90453-5

[42] S. Shah, A. Sharma, M.G.-A. Biochemistry, U. 2006, Preparation of cross-linked enzyme aggregates by using bovine serum albumin as a proteic feeder, Elsevier. (2006). https://www.sciencedirect.com/science/article/pii/S0003269706000649?casa_token=y ks_kKd2FzsAAAAA:eshqtuZMzRAlJkVKwNc6Ad9sbbTnmNlO8l-Q2CWTK9noUT1belSRIxr5fTvHFpiWR_M9Pqf6wU6l (accessed August 14, 2021).

[43] F. Šulek, D. Fernández, Ž. Knez, … M.H.-P., U. 2011, Immobilization of horseradish peroxidase as crosslinked enzyme aggregates (CLEAs), Elsevier. (2011). https://www.sciencedirect.com/science/article/pii/S1359511310004526?casa_token=-R3nyT8sdEYAAAAA:8cqC2aerKNG5c2e8lWcvYE6D36cKlLtnPJHHQQTt6gi4dG9 GQKMK5bLx8klPwbI1SBqkPV3tI4s4 (accessed August 14, 2021).

[44] J. Chen, J. Zhang, B. Han, Z. Li, J. Li, X.F.S.B. Biointerfaces, U. 2006, Synthesis of cross-linked enzyme aggregates (CLEAs) in CO2-expanded micellar solutions, Elsevier. (2006). https://www.sciencedirect.com/science/article/pii/S0927776506000361?casa_token=0 WVjHRhMYYAAAAAA:EojZlOloTj2TantifJmqYToH_euNym22EAZ1H-6OediuTCxbC5Xs_garAyjnRCulnZij9Ck20B9f (accessed August 14, 2021).

[45] N. Chauhan, A. Kumar, C.S. Pundir, Construction of an Uricase Nanoparticles Modified Au Electrode for Amperometric Determination of Uric Acid, Appl. Biochem. Biotechnol. 174 (2014) 1683-1694. https://doi.org/10.1007/s12010-014-1097-6

[46] L. Cao, F. Van Rantwijk, R.A. Sheldon, Cross-linked enzyme aggregates: A simple and effective method for the immobilization of penicillin acylase, Org. Lett. 2 (2000) 1361-1364. https://doi.org/10.1021/ol005593x

[47] S. Chawla, R. Rawal, Sonia, Ramrati, C.S. Pundir, Preparation of cholesterol oxidase nanoparticles and their application in amperometric determination of cholesterol, J. Nanoparticle Res. 15 (2013). https://doi.org/10.1007/s11051-013-1934-5

[48] M. del P. Guauque Torres, M.L. Foresti, M.L. Ferreira, Cross-linked enzyme aggregates (CLEAs) of selected lipases: A procedure for the proper calculation of their recovered activity, AMB Express. 3 (2013) 1-11. https://doi.org/10.1186/2191-0855-3-25

[49] J.S. Beck, J.C. Vartuli, W.J. Roth, M.E. Leonowicz, C.T. Kresge, K.D. Schmitt, C.T.W. Chu, D.H. Olson, E.W. Sheppard, S.B. McCullen, J.B. Higgins, J.L. Schlenker, A New Family of Mesoporous Molecular Sieves Prepared with Liquid Crystal Templates, J. Am. Chem. Soc. 114 (1992) 10834-10843. https://doi.org/10.1021/ja00053a020

[50] E.M. -, Immobilisation of enzymes on mesoporous silicate materials, Pubs.Rsc.Org. (2013). https://doi.org/10.1039/c0xx00000x.

[51] Z. Zhou, M.H.-C.S. Reviews, U. 2013, Progress in enzyme immobilization in ordered mesoporous materials and related applications, Pubs.Rsc.Org. (2013).

https://pubs.rsc.org/en/content/articlehtml/2013/cs/c3cs60059a (accessed August 19, 2021).

[52] H.L. Betancor, L, Bioinspired enzyme encapsulation for biocatalysis, Elsevier. (2008). https://www.sciencedirect.com/science/article/pii/S0167779908001972?casa_token=G ickHzUafdwAAAAA:xcxwvecedu6drzMbjitcgrb88Ed8NWJ8FSY5htk9nVul9A9aGct RjbJAvYsKpoIrkyr8h40M2A8Y (accessed August 19, 2021).

[53] E. Cipolatti, A. Valerio, R. Henriques, … D.M.-, Nanomaterials for biocatalyst immobilization-state of the art and future trends, Pubs.Rsc.Org. (2016). https://pubs.rsc.org/en/content/articlehtml/2016/ra/c6ra22047a (accessed August 19, 2021). https://doi.org/10.1039/C6RA22047A

[54] W. Wei, J. Du, J. Li, M. Yan, Q. Zhu, X. Jin, … X.Z.-A., undefined 2013, Construction of robust enzyme nanocapsules for effective organophosphate decontamination, detoxification, and protection, Wiley Online Libr. 25 (2013) 2212-2218. https://doi.org/10.1002/adma.201205138

[55] X. Wu, M. Hou, J.G.-C.S.& Technology, U. 2015, Metal-organic frameworks and inorganic nanoflowers: a type of emerging inorganic crystal nanocarrier for enzyme immobilization, Pubs.Rsc.Org. (2015). https://pubs.rsc.org/en/content/articlehtml/2015/cy/c5cy01181g (accessed August 19, 2021).

[56] F. Lyu, Y. Zhang, R. Zare, J. Ge, Z.L.-N. letters, undefined 2014, One-pot synthesis of protein-embedded metal-organic frameworks with enhanced biological activities, ACS Publ. 14 (2014) 5761-5765. https://doi.org/10.1021/nl5026419

[57] J. Ge, J. Lei, R.Z.-N. nanotechnology, undefined 2012, Protein-inorganic hybrid nanoflowers, Nature.Com. (2012). https://doi.org/10.1038/NNANO.2012.80. https://doi.org/10.1038/nnano.2012.80

[58] I. Ocsoy, E. Dogru, S.U.-E. and microbial Technology, U. 2015, A new generation of flowerlike horseradish peroxides as a nanobiocatalyst for superior enzymatic activity, Elsevier. (2015). https://www.sciencedirect.com/science/article/pii/S0141022915000691?casa_token=O 4H31opcXIAAAAA:Jr8vAt4s7DoT3rjKT8GikatF3HSHYi0Yc0fDevKbbxu5yf1ZhJhs zuAM9sC2jA0hHryiH3NoEfpI (accessed August 18, 2021).

[59] T.. A. Pradeep, Noble metal nanoparticles for water purification: a critical review, Elsevier. (2009). https://www.sciencedirect.com/science/article/pii/S004060900900683X?casa_token=t

gny1ABytbkAAAAA:JgP4aX6syJ3c1zli5_niSWd97x9VDFwv-
5EaZF1ia7GvRz_d2U2q0s9v2BKxNxNQ0HP2pWp0wEDU (accessed August 19,
2021).

[60] J. Liu, Y. Lu, A colorimetric lead biosensor using DNAzyme-directed assembly of
gold nanoparticles, J. Am. Chem. Soc. 125 (2003) 6642-6643.
https://doi.org/10.1021/ja034775u

[61] L. Stadlmair, T. Letzel, J. Drewes, J.G.- Chemosphere, U. 2018, Enzymes in
removal of pharmaceuticals from wastewater: A critical review of challenges,
applications and screening methods for their selection, Elsevier. (2018).
https://www.sciencedirect.com/science/article/pii/S0045653518307914?casa_token=S
RptH95Y8J0AAAAA:kt8Rrf7B7a4RA7vl7o5Druj9AQF7Yu4MfgQxlC_RvAZPr8Kd
KXlU42c9rzbhZ4m2TpUbrTjIANYR (accessed August 18, 2021).

[62] B. Sharma, A. Dangi, P.S.-J. of environmental Management, U. 2018, Contemporary
enzyme based technologies for bioremediation: a review, Elsevier. (2018).
https://www.sciencedirect.com/science/article/pii/S030147971731263X?casa_token=e
HJisPsz22EAAAAA:_erS5fYdquf6UXC3TYLV6EihpGSPLt5-
1TvksC6u6gP65jkuE6ZWAMZ7QbdGJon_ZQqpUTZL9aZy (accessed August 18,
2021).

[63] M. Wang, Y. Chen, V.A. Kickhoefer, L.H. Rome, P. Allard, S. Mahendra, A Vault-
Encapsulated Enzyme Approach for Efficient Degradation and Detoxification of
Bisphenol A and Its Analogues, ACS Sustain. Chem. Eng. 7 (2019) 5808-5817.
https://doi.org/10.1021/acssuschemeng.8b05432

[64] M. Bilal, M. Adeel, T. Rasheed, Y.Z.-E. International, U. 2019, Emerging
contaminants of high concern and their enzyme-assisted biodegradation-a review,
Elsevier. (2019).
https://www.sciencedirect.com/science/article/pii/S0160412018323523 (accessed
August 18, 2021).

[65] C.A. Gasser, L. Yu, J. Svojitka, T. Wintgens, E.M. Ammann, P. Shahgaldian, P.F.X.
Corvini, G. Hommes, Advanced enzymatic elimination of phenolic contaminants in
wastewater: A nano approach at field scale, Appl. Microbiol. Biotechnol. 98 (2014)
3305-3316. https://doi.org/10.1007/s00253-013-5414-8

[66] W. Wei, J. Du, J. Li, M. Yan, Q. Zhu, X. Jin, … X.Z.-A., undefined 2013,
Construction of robust enzyme nanocapsules for effective organophosphate
decontamination, detoxification, and protection, Wiley Online Libr. 25 (2013) 2212-
2218. https://doi.org/10.1002/adma.201205138

[67] X. Fang, C. Zhang, X. Qian, D.Y.-R. Advances, U. 2018, Self-assembled 2, 4-dichlorophenol hydroxylase-inorganic hybrid nanoflowers with enhanced activity and stability, Pubs.Rsc.Org. (2018). https://pubs.rsc.org/en/content/articlehtml/2018/ra/c8ra02360c (accessed August 18, 2021). https://doi.org/10.1039/C8RA02360C

[68] Y. Azuma, D.L.V. Bader, D. Hilvert, Substrate Sorting by a Supercharged Nanoreactor, J. Am. Chem. Soc. 140 (2018) 860-863. https://doi.org/10.1021/jacs.7b11210

[69] D.C. Buehler, M.D. Marsden, S. Shen, D.B. Toso, X. Wu, J.A. Loo, H. Zhou, V.A. Kickhoefer, P.A. Wender, J.A. Zack, L.H. Rome, D.C. Buehler, M.D. Marsden, Bioengineered vaults: Self-assembling protein shell-lipophilic core nanoparticles for drug delivery, ACS Publ. 8 (2014) 7723-7732. https://doi.org/10.1021/nn5002694

[70] P. Dvorak, S. Bidmanova, J.D.-... science & technology, undefined 2014, Immobilized synthetic pathway for biodegradation of toxic recalcitrant pollutant 1, 2, 3-trichloropropane, ACS Publ. 48 (2014) 6859-6866. https://doi.org/10.1021/es500396r

[71] M. Wang, S.K. Mohanty, S. Mahendra, Nanomaterial-Supported Enzymes for Water Purification and Monitoring in Point-of-Use Water Supply Systems, Acc. Chem. Res. 52 (2019) 876-885. https://doi.org/10.1021/acs.accounts.8b00613

[72] B. Thallinger, E.N. Prasetyo, G.S. Nyanhongo, G.M. Guebitz, Antimicrobial enzymes: An emerging strategy to fight microbes and microbial biofilms, Biotechnol. J. 8 (2013) 97-109. https://doi.org/10.1002/biot.201200313

[73] H. Luckarift, M. Dickerson, K. Sandhage, J.S.- Small, undefined 2006, Rapid, room-temperature synthesis of antibacterial bionanocomposites of lysozyme with amorphous silica or titania, Wiley Online Libr. 2 (2006) 640-643. https://doi.org/10.1002/smll.200500376

[74] R.C. Pangule, S.J. Brooks, C.Z. Dinu, S.S. Bale, S.L. Salmon, G. Zhu, D.W. Metzger, R.S. Kane, J.S. Dordick, Antistaphylococcal nanocomposite films based on enzyme - Nanotube conjugates, ACS Nano. 4 (2010) 3993-4000. https://doi.org/10.1021/nn100932t

[75] D. Patterson, K. McCoy, ... C.F.-J. of M., U. 2014, Constructing catalytic antimicrobial nanoparticles by encapsulation of hydrogen peroxide producing enzyme inside the P22 VLP, Pubs.Rsc.Org. (2014). https://pubs.rsc.org/en/content/articlehtml/2014/tb/c4tb00983e (accessed August 18, 2021).

[76] S. Bang, A. Jang, J. Yoon, P. Kim, … J.K.-E. and microbial, U. 2011, Evaluation of whole lysosomal enzymes directly immobilized on titanium (IV) oxide used in the development of antimicrobial agents, Elsevier. (2011). https://www.sciencedirect.com/science/article/pii/S0141022911001207?casa_token=a mgyPERlWpEAAAAA:TV1NwjuB9NvaEuf65YK5V9cYERVhmZeoKVtI90A3GY9 FanRHkOxGKOxAxbMsNRzdQVUD-8DjK5aI (accessed August 18, 2021).

[77] P. Asuri, S. Karajanagi, R. Kane, J.D.- Small, undefined 2007, Polymer-nanotube-enzyme composites as active antifouling films, Wiley Online Libr. 3 (2007) 50-53. https://doi.org/10.1002/smll.200600312

[78] X. Qu, J. Brame, Q. Li, P.A.-A. of chemical research, undefined 2013, Nanotechnology for a safe and sustainable water supply: enabling integrated water treatment and reuse, ACS Publ. 28 (2012) 834-843. https://doi.org/10.1021/ar300029v

[79] R. Deo, J. Wang, I. Block, A. Mulchandani, … K.J.-A. chimica, U. 2005, Determination of organophosphate pesticides at a carbon nanotube/organophosphorus hydrolase electrochemical biosensor, Elsevier. (2005). https://www.sciencedirect.com/science/article/pii/S0003267004012875?casa_token=x 91ikJELhUkAAAAA:68r6gcaanwaGV31mEAE8m9AmMnauuSpu62l8Ix5u2_z5Z0R 8-EfFRuCinmmFttBN40ueahivDg5k (accessed August 18, 2021).

[80] Y. Li, Z. Liu, Y. Liu, Y. Yang, G. Shen, R.Y.-A. Biochemistry, U. 2006, A mediator-free phenol biosensor based on immobilizing tyrosinase to ZnO nanoparticles, Elsevier. (2006). https://www.sciencedirect.com/science/article/pii/S0003269705008274?casa_token=I KEGHLZ3H9EAAAAA:r4er0HTxjmxLWL_lIIiD3Uxe263CvJepZGzhKvIGWFqpc3r YPxTKiwld5gpYIQGVrd43K4wtKCG5 (accessed August 18, 2021).

[81] D. Pan, Y. Gu, H. Lan, Y. Sun, H.G.-A. chimica Acta, U. 2015, Functional graphene-gold nano-composite fabricated electrochemical biosensor for direct and rapid detection of bisphenol A, Elsevier. (2015). https://www.sciencedirect.com/science/article/pii/S0003267014013178?casa_token=G UFZI6KPhN4AAAAA:ssktMX-00yb3iEkqEKERg4FYezsn2IIsCbCzb4sjxUes4z8- RzPK89X0GtvyjhUv1GwiWjA8DD11 (accessed August 18, 2021).

[82] G. Liu, Y. Lin, Electrochemical sensor for organophosphate pesticides and nerve agents using zirconia nanoparticles as selective sorbents, Anal. Chem. 77 (2005) 5894-5901. https://doi.org/10.1021/ac050791t

[83] A. Amine, H. Mohammadi, … I.B.-B. and, U. 2006, Enzyme inhibition-based biosensors for food safety and environmental monitoring, Elsevier. (2006).

https://www.sciencedirect.com/science/article/pii/S0956566305002198?casa_token=c
0mIQlLs_HUAAAAA:sC0cEbCPmv5aJOqz10fKexEewEJukIaRuDOfqy3DjvfLtCufd
hFJQpAp6lackNYn373zWosSbplD (accessed August 18, 2021).

[84] Y. Yang, Z. Wang, M. Yang, M. Guo, Z. Wu, … G.S.-S. and A.B., U. 2006,
Inhibitive determination of mercury ion using a renewable urea biosensor based on
self-assembled gold nanoparticles, Elsevier. (2006).
https://www.sciencedirect.com/science/article/pii/S0925400505004065?casa_token=jn
1NnRK164sAAAAA:1R_0gbci9_iIm4G5RZlNte_P_2_FS4ZgdbCwPxOSSSim79Jdn
AQvgf5HBPcoDcb3HY7PklDzAphO (accessed August 19, 2021).

[85] L. Zhu, L. Gong, Y. Zhang, R. Wang, … J.G.-C.A., undefined 2013, Rapid detection
of phenol using a membrane containing laccase nanoflowers, Wiley Online Libr. 8
(2013) 2358-2360. https://doi.org/10.1002/asia.201300020

[86] O.R. Miranda, X. Li, L. Garcia-Gonzalez, Z.J. Zhu, B. Yan, U.H.F. Bunz, V.M.
Rotello, Colorimetric bacteria sensing using a supramolecular enzyme-nanoparticle
biosensor, J. Am. Chem. Soc. 133 (2011) 9650-9653.
https://doi.org/10.1021/ja2021729

[87] P.K. Kumar, V., Singh, D., Sangwan, P., and Gill, Industrial enzymes : trends, scope
and relevance, New York Nov. Sci. Publ. (2014) 211.

[88] R. Pang, M. Li, C.Z.- Talanta, U. 2015, Degradation of phenolic compounds by
laccase immobilized on carbon nanomaterials: diffusional limitation investigation,
Elsevier. (2015).
https://www.sciencedirect.com/science/article/pii/S0039914014006134?casa_token=6-
wrDwPCSaYAAAAA:6Nx46qbB4yg3GYXaNE017oGku5qgUcjsD6uFpSdXcUUN6
wc79TdmHuoVCFybv29zgIxwmGLRrLW2 (accessed August 22, 2021).

[89] A. Simonian, T. Good, S. Wang, J.W.-A. chimica Acta, U. 2005, Nanoparticle-based
optical biosensors for the direct detection of organophosphate chemical warfare agents
and pesticides, Elsevier. (2005).
https://www.sciencedirect.com/science/article/pii/S0003267004008050?casa_token=-
kMFOk1zuckAAAAA:tmMUtvAQEIRR8t8UEngvFwsblHQkd_GacMSGLUviKkuYz
g3MC1rvTwgbDkm58vAbNylXsDfVkmry (accessed August 18, 2021).

Nanomaterial-Supported Enzymes   Materials Research Forum LLC
Materials Research Foundations **126** (2022) 142-161   https://doi.org/10.21741/9781644901977-5

Chapter 5

Enzyme Immobilized Nanoparticles Towards Biosensor Fabrication

Jaison Jeevanandam[1], Sharadwata Pan[2], Michael K. Danquah[3]*

[1]CQM-Centro de Química da Madeira, MMRG, Universidade da Madeira, Campus da Penteada, 9020-105 Funchal, Portugal

[2]TUM School of Life Sciences, Technical University of Munich, 85354 Freising, Germany

[3]Chemical Engineering Department, University of Tennessee, Chattanooga, 615 McCallie Ave, Chattanooga, TN 37403, United States

* michael-danquah@utc.edu

Abstract

In recent times, nanomaterials with semiconductor properties are introduced as a potential transducer in biosensors, which can be credited to their intrinsic, elevated surface-to-volume proportion, enhanced sensitivity, and improved surface properties. The surface properties of nanomaterials have made them a significant transducer matrix towards the immobilization of bioreceptors, which eventually enhances the identification threshold and the biosensor sensing capability. Several nanomaterials, such as polymer, metal oxide, metal and carbon-based, as well as nanocomposites, are used towards transducer manufacturing, eventually being incorporated in the biosensors. The current chapter lays an outline with respect to biosensors that are fabricated with nanomaterials as a transducer, where enzymes acting as a bioreceptor, are immobilized on their surface. In addition, the biosensing mechanisms of the enzyme immobilized nanomaterials, their efficiency, detection limit, and sensitivity, are also discussed.

Keywords

Biosensor, Enzyme, Immobilization, Nanoparticles, Biomolecules, Nanocomposites

Contents

1.   Introduction

Sensors are electronic or electric devices, which are beneficial towards the detection or monitoring of an analyte of interest [1]. The sensors, which utilize biomolecules as an agent to detect biological compounds or biomarkers, are termed biosensors [2]. In general, a biosensor will possess some basic parts, such as a bio-receptor, which could be an enzyme, an antibody or a nucleic acid, semiconductor or nanomaterial as a transducer, and an electronic system, such as a processor, a display, and a signal amplifier [3]. The biosensors are widely classified into electrochemical [4], optical [5], electronic [6], piezoelectric [7], gravimetric [8], pyroelectric [9], and magnetic biosensors [10], depending on the type of bio-receptor used in the biosensing detection system. These biosensors possess several advantages, such as high selectivity, sensitivity, as well as high-throughput process abilities [11]. However, conventional semiconductor materials as a transducer possess limitations, such as less sensitivity and the inability to detect multiple bio-compounds [12]. Thus, there is a perpetual thrust towards the design and inception of a novel alternative material to be used as a transducer in biosensors.

In recent times, nanomaterials with semiconductor properties are introduced as a potential transducer in biosensors, which can be credited to their intrinsic, elevated surface-to-volume proportion, enhanced sensitivity, and improved surface properties [13-16]. The surface properties of nanomaterials have made them a significant transducer matrix towards the immobilization of bioreceptors, which eventually enhances the identification threshold and the biosensor sensing capability [17]. Several nanomaterials, such as polymer, metal oxide, metal and carbon-based, as well as nanocomposites, are utilized towards transducer synthesis, eventually being assimilated in the biosensors [18]. The current chapter lays an outline with respect to biosensors, which are manufactured with nanomaterials as a transducer, where enzymes as bioreceptor are immobilized on their surface. In addition, the biosensing mechanisms of the enzyme immobilized nanomaterials, their efficiency, detection limit, and sensitivity, are also discussed.

2.   Enzyme immobilized nanomaterials

Nanomaterials, which are made up of metal, metal oxide, polymer, carbon-based nanoparticles, and nanocomposites, are widely used to immobilize enzymes to be utilized as a transducer for the detection of biomolecules via bioreceptors [19]. This section emphasizes various nanomaterial types as a matrix that is used to immobilize enzymes towards biomedical applications.

2.1   Metal nanomaterials

Gold nanoparticles are widely used in biosensor applications, and their efficiency is improved by enzyme immobilization, as shown in Figure 1. Recently, electrodes (gold screen-printed) crusted with bi-layered hydroxides (made of aluminum and cobalt) were utilized for the fabrication of biosensors by immobilizing tyrosinase on the surface of the nanomaterial transducer matrix [20]. Likewise, Majouga et al. (2015) fabricated novel, gold-coated magnetite-based core-shell nanoparticles that are functionalized with alpha-chymotrypsin enzyme via immobilization. The study showed that the catalytic action of the immobilized enzyme can be affected (decreased) by a low frequency alternating current (AC) magnetic field, due to the magnetic properties of the nanomaterial [21]. Similarly, Lu et al. (2007) fabricated novel gold nanowires via electrodeposition approach, using a membrane of nanosized polycarbonate pore. Further, the glucose oxidase enzyme was immobilized on the gold nanowire surfaces for the synthesis of the glucose biosensor [22]. Later, Du et al. (2011) fabricated gold nanorods as a nanosized carrier for the horseradish peroxidase (HRP) enzyme via a multienzyme amplification approach. This enzyme immobilized gold nanorods were identified to be beneficial for the detection of Ab2 antibodies via thionine (oxidized by peroxidase) reduction, with the presence of hydrogen peroxide [23]. Moreover, silver nanoparticles were synthesized and were made to interact with metalized nucleotides and adenosine triphosphate. These enzyme-responsive silver nanoparticles were demonstrated to possess an enhanced anticancer activity against the human liver carcinoma cells, as well as countering actions against both gram-negative and gram-positive bacilli [24]. Furthermore, platinum [25], copper [26], and other rare earth metal nanoparticles [27] were used as a matrix for the enzymes to be immobilized on their surfaces, towards the fabrication of an efficient biosensor.

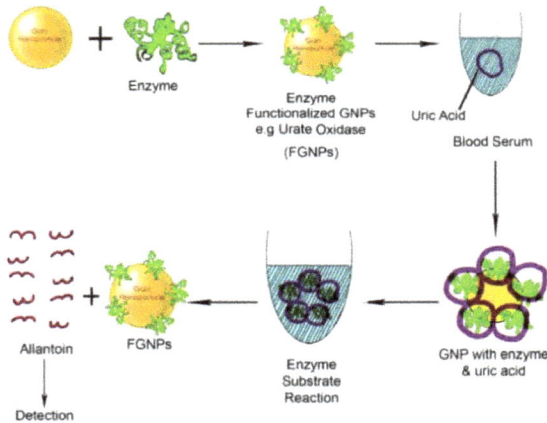

Figure 1. Gold nanoparticles functionalized with enzyme via immobilization. Reproduced with permission from [28], © MDPI, 2011 [open access].

2.2   Metal oxide nanomaterials

Apart from metal nanoparticles, metal oxide nanoparticles were also exploited towards biosensor fabrication or other applications, where the enzymes were used to functionalize them on the surface via an immobilization process, as shown in Figure 2. Rani et al. (2017) recently immobilized laccase enzyme on zinc oxide and manganese dioxide nanomaterial (eco-synthesized) surfaces. The enzyme immobilized nanomaterial was used to advance the enzyme catalyzing feature, aimed towards the degradation of a harmful dye (alizarin red S) [29]. Further, Singh et al. (2014) prepared novel thulium oxide nanorod structures via the hydrothermal approach, where cholesterol oxidase and esterase enzymes were immobilized on their surfaces towards the fabrication of potential biosensors. The enzyme immobilized metal oxide nanomaterial was accumulated (electrophoretically) on the indium-tin-oxide glass substrate surface, towards the quantification of total cholesterol in serum (human) specimens, with a response time of 40 s [30]. Furthermore, Kant and Gupta

(2018) synthesized tantalum oxide nanoflakes with multilayers of the silver metal, which were functionalized with the acetylcholinesterase enzyme via immobilization. The fabricated enzyme immobilized rare earth metal oxide was used for the development of a fiber optic surface plasmon resonance (SPR)-based acetylcholine biosensor [31]. Moreover, Verma et al. (2017) fabricated zinc oxide nanoparticles and immobilized cysteine enzymes towards enhanced biosensor applications. The study showed that the zinc oxide nanoparticles improve the enzyme loading by up to 62.9% with the enzyme-specific activity of 72.45% [32].

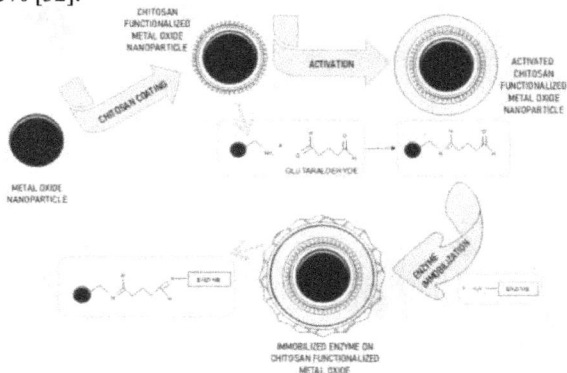

Figure 2. Metal oxide magnetic nanoparticles functionalized with enzyme via immobilization towards therapeutic applications. Reproduced with permission from [33], © MDPI, 2020 [open access].

2.3   Carbon-derived nanomaterials

Recently, carbon-based nanomaterials, for instance graphene (see Figure 3), fullerene, carbon nanotubes (CNTs) and carbon dots, are also used as a matrix towards the immobilization of enzymes, eventually targeting biosensor applications. Wang et al. (2017) fabricated a multilayered gold nanoparticle-graphene-titanium dioxide nanocomposite, that is functionalized with horseradish peroxidase enzyme. These novel enzyme immobilized graphene-based biosensors were reported to be advantageous towards an enhanced identification of thrombin in serum (human) specimens [34]. Likewise, Lai et al. (2016) synthesized a graphene oxide (GO) nanoprobe that is functionalized with the HRP enzyme via immobilization. The resultant nanomaterial was utilized as a biosensor carrying out an ultrasensitive immunoassay towards enhanced ferrocene quantification [35]. Similarly, Zhou et al. (2020) prepared novel C60 fullerene nanoparticles that are functionalized with a duplex-specific nuclease enzyme via the click-chemistry mediated enzyme-assisted target recycling approach. The resultant nanomaterial was used towards the detection of microRNA-141 via a dual-amplified strategy-based ultrasensitive electrochemical biosensor [36]. Besides, Afreen et al. (2015) proposed that fullerenes functionalized with enzymes, particularly myoglobin, urease and glucose oxidase, can be highly beneficial as nanomediators, towards the manufacturing of delicate biosensors [37]. Further, Su et al. (2018) immobilized HRP on the surface of carbon dots (amine-functionalized) through a single-step hydrothermal methodology. The synthesized nanomaterial on the electrode (glassy carbon) was used towards a sensitive hydrogen peroxide detection [38]. Furthermore, Goncalves et al. (2010) fabricated novel carbon dots via direct laser ablation, which were functionalized with N-acetyl-L-cysteine and amine-polyethylene glycol. The functionalized carbon dot could be utilized towards the detection and measurement of mercury via the fluorescence property of the nanomaterial [39]. Meanwhile, NADH oxidase was surface-immobilized on functionalized CNTs via a specific and reversible immobilization process [40]. In addition, HRP was surface-immobilized on silica nanoparticles, and were mixed with an aqueous ink prepared using single-walled CNTs. The novel nanomaterial formulation was utilized towards the fabrication of fully inkjet-printed biosensors, in order to detect hexacyanoferrate ions or hydroquinone [41].

2.4   Polymeric nanomaterials

In recent times, polymer-based nanomaterials are also gaining significance to immobilize enzymes on their surface for biosensor applications. Yoon et al. (2008) demonstrated that novel carboxylated polypyrrole nanotubes can be synthesized, where pyrrole-3-carboxylic acid can be used to functionalize their surface to bind the glucose oxidase enzyme via immobilization. The study showed that the nanomaterial can be utilized as a potential field-effect-transistor sensor for monitoring glucose [43]. Further, Keller et al. (2017) showed that the 'core-shell' (polymeric) nanoparticles, functionalized with nitrilotriacetic amine, can be utilized to sustain enzyme immobilization. In this work, the nitroxide-mediated polymerization of nitroxide SG1 was used to protect the amine moiety via a self-assembly approach, induced by polymerization as immobilization support for polyhistidine-functionalized HRP enzyme [44]. Recently, Bezerra et al. (2020) fabricated iron oxide magnetic nanoparticles, which were activated with divinyl sulfone, and functionalized with polyethylene amine. The resultant nanomaterial was identified to be beneficial towards the immobilization of lipase, extracted from the *Thermomyces lanuginosus* bacteria [45].

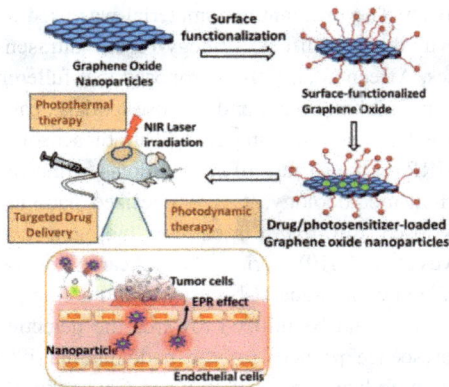

Figure 3. Graphene oxide nanoparticles functionalized with enzyme via immobilization towards therapeutic applications. Reproduced with permission from [42], © MDPI, 2020 [open access].

2.5 Nanocomposites

It is noteworthy that in recent times nanocomposites are increasingly being utilized to immobilize enzymes on their surface towards enhanced biosensor applications. Jun et al. (2019) immobilized the Jicama peroxidase enzyme on the surface of a nanocomposite membrane, which is made up of buckypaper, CNTs (multi-walled), and PVA (polyvinyl alcohol). The study exhibited that the enzyme immobilized on the surface of nanocomposite displayed a significant enhancement in their pH, storage, and thermal stabilities, compared to the free enzyme [46]. Further, Asmat et al. (2017) fabricated a nanocomposite with GO, which was functionalized with silver, and coated with polyaniline, towards the immobilization of the lipase enzyme, extracted from the fungus *Aspergillus niger*. The resultant enzyme immobilized nanocomposite was revealed to possess around 94% of immobilization yield, and 88.5% of high enzymatic activity, along with the enhanced structural rigidity of the enzyme [47]. Furthermore, Vineh et al. (2020) recently showed that the nanocomposite, synthesized using reduced graphene oxide (rGO)-silicon dioxide, possesses the ability to be functionalized by HRP via a covalent bond-based immobilization process. The resultant nanocomposite was proposed to be beneficial towards the biodegradation of phenols and a wide range of dyes, including methyl green, phenol red, methyl red, methyl orange, reactive black 5, Coomassie brilliant blue, bromocresol green, bromophenol blue, and bromothymol blue [48].

Nanomaterial-Supported Enzymes   Materials Research Forum LLC
Materials Research Foundations **126** (2022) 142-161   https://doi.org/10.21741/9781644901977-5

3.   Enzyme immobilized nanomaterial-based biosensors and their applications

The enzyme immobilized nanomaterials are widely exploited towards the manufacturing of electrochemical, optical, piezoelectric, gravimetric and magnetic biosensors in recent times as shown in Table 1.

Table 1*. Applications of enzyme immobilized nanomaterial-based biosensors*

Nanomaterial	Immobilized enzyme	Application	Reference
Multiwalled carbon nanotube	Bovine serum albumin	Detect phosphate in the artificial saliva	[49]
Multiwalled carbon nanotube with zinc oxide nanoparticles	Laccase	Bisphenol A detection	[50]
Graphene oxide nanoparticles	Glucose oxidase	Glucose sensor	[51]
Gold nanoparticle and graphene oxide	Uricase enzyme	Uric acid quantification	[53]
Zeolitic imidazolate framework	Urease	Urea quantification	[54]
Tantalum oxide nanoflower	Acetylcholinesterase	Acetylcholine Detection	[55]
Platinum and cadmium selenide nanoparticles	Tyrosine methylester	Tyrosinase quantification	[57]
Iron oxide-silicon dioxide core-shell magnetic nanoparticles	Glucose oxidase	Glucose sensor	[60]
Magnetic ferrite nanoparticles in carbon nanotube	Horseradish peroxidase and cholesterol oxidase	Cholesterol detection	[61]
Iron oxide nanoparticles	Beta-glucosidase	Identification of beta-glucosidase enzyme	[62]

3.1   Electrochemical biosensors

Figure 4 is the schematic representation of enzyme immobilized nanomaterials for manufacturing nanosensors. Recently, Bai et al. (2021) employed CNTs (multi-walled) as an enzyme carrier, where bovine serum albumin (BSA) and glutaraldehyde were immobilized on their surface via a printing strategy (layer-by-layer) via the interconnection of the ink constituents amongst the strata of printed constituents. In this study, the working electrode was preloaded with the reagents for the catalytic enzyme reaction, which was coupled with a smartphone (Android) application for real-time processing of the data. The fabricated biosensor system showed a fast reaction time (< 10 seconds) for detecting phosphate in the artificial saliva sample (30 µl), with a high stability and selectivity towards efficient, point-of-care applications [49]. Similarly, Kunene et al. (2018) modified a printed (carbon-screen) electrode with CNTs (multi-walled) functionalized with nanoparticles (zinc oxide), which were doped with silver, where the laccase enzyme was bound on their surface via immobilization. The resultant enzyme immobilized nanocomposite-based biosensor showed an enhanced ability for quantifying Bisphenol A, inclduing a $0.5 - 2.99$ µM linear range performance, a 6 nM detection limit, and a 0.86% high reproducible response factor, in plastic bottle samples [50]. Likewise, Akhtar et al. (2019) manufactured a GO on an interface (screen-printed, gold-sputtered) surface, where the glucose oxidase immobilization was expedited by the 1-ethyl-3-(3-(dimethylamino)propyl) carbodiimide. The resultant biosensor was identified to possess a larger sensing ability magnitude of 3.1732 µA mM^{-1} cm^{-2} towards glucose, including a 3-9 mM linear response range, and a 0.3194 mM detection limit [51].

3.2   Optical biosensors

Singh et al. (2019) fabricated a novel composite with gold nanoparticles and GO, functionalized with pointed fiber, and uricase enzyme on their surface, via immobilization. The resultant nanomaterial was identified to possess an enhanced, localized SPR property for uric acid quantification. The study showed that the optical biosensor possesses the ability to detect distinct uric acid proportions between 10-800 µM, where their selectivity is improved via the immobilization of the uricase enzyme [53]. Besides, Zhu et al. (2019) demonstrated the fabrication of a novel enzyme capsulation film towards optical biosensor applications, with the zeolitic imidazolate framework (few nanometers in size) embedded with the urease enzyme via *in situ* development procedure, on the fiber (devoid of core) surface. The fabricated novel film exhibited an excellent linear relationship for quantifying urea (1 to 10 mM range in concentration) in real samples, including a quantification threshold of 0.1 mM within a scope of 1525–1590 nm of broadband light (0.8 mM/refractive index unit) [54]. Meanwhile, Kant and Gupta (2019) prepared an optical biosensor founded on SPR aimed towards quantifying acetylcholine via its counterpart enzyme, which is entrapped via immobilization on a tantalum oxide nanoflower assembly that are encapsulated in the matrix of rGO and chitosan. The results showed that the optical biosensor detects acetylcholine within $0 - 8$ µM of concentration range, with 73 nM of qunatification threshold [55].

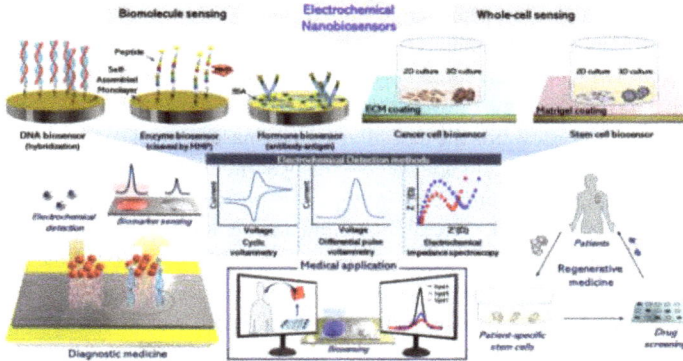

Figure 4. Enzyme immobilized nanoparticles towards electrochemical nanosensor applications. Reproduced with permission from [52], © MDPI, 2020 [open access].

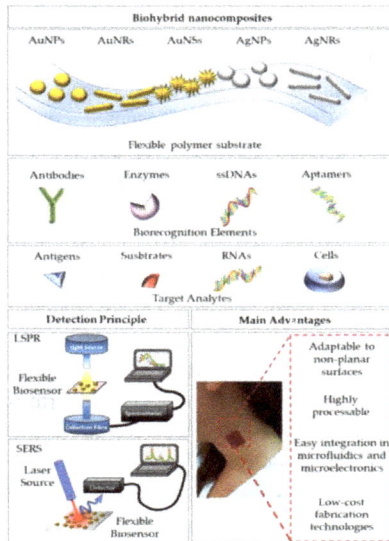

Figure 5. Enzyme immobilized nanoparticles towards optical biosensor applications. Reproduced with permission from [56], © MDPI, 2020 [open access].

3.3   Piezoelectric and gravimetric biosensor

Piezoelectric and gravimetric biosensors embody the most infrequent biosensing devices, which exploit enzyme immobilized nanomaterials, as the detection process may lead to heat generation, which may disintegrate the enzymes. Yildiz et al. (2008) modified the surface of platinum and cadmium selenide nanoparticles with tyrosine methyl ester for quantifying tyrosinase, which represents a symptomatic indicator of cancerous cells in the skin (melanoma). The study showed that the nanomaterial can be used as a potential piezoelectric biosensor towards the detection of tyrosinase with 1 U and 0.1 U of detection limits, via electrochemical and photoelectrochemical methods, respectively [57]. Further, the gravimetric biosensor embodies a type of piezoelectric device, in which the biomolecules can be detected, thereby provoking an alteration in the resonance frequency, corresponding to the analyte mass [58]. Thus, Jia et al. (2021) integrated the reverberating nano or micro sensors (gravimetric) towards the biochemical recognition of biomolecules in liquid and air. The gravimetric sensor with immobilized enzymes was proposed to be beneficial towards the detection of trace analytes in bacteria, biomarkers, DNA, volatile organic compounds, chemical vapors, and pollutant gases, via distinct in-plane, torsional, extensional, and flexural modes [59].

3.4   Magnetic biosensors

Qiu et al. (2007) demonstrated the fabrication of iron oxide-silicon dioxide core-shell magnetic nanoparticles, which are modified with ferrocene, and can be used as building blocks for the reagent-less enzyme-based biosensor construction. In this study, the biosensor was fabricated via the entrapment of glucose oxidase in the composite of chitosan, doped with core-shell nanoparticles, and modified with ferrocene monocarboxylic acid. The resultant biosensor was identified to possess a linear glucose detection range of 1.0×10^{-5} to 4.0×10^{-3} M, with a 3.2 µM of detection limit [60]. Likewise, Eguilaz et al. (2011) designed a magnetic high-performance bienzyme biosensor with the help of magnetic nanoparticles, which are wrapped in the platform of CNTs. In this study, nanoparticles (ferrite, magnetic) were subjected to functionalization via the incorporation of poly (diallyl dimethylammonium chloride) and glutaraldehyde, which were coated with CNTs (multi-walled). This novel platform was immobilized with HRP and cholesterol oxidase enzymes on their surface, for fabricating bienzyme biosensors. The resultant biosensor could identify cholesterol in the concentration range of 0.01 – 0.95 mM at -0.05 V of applied potential, with 0.85 µM of detection limit and 1.57 mM of Michaelis-Menten constant [61]. Recently, Zhang et al. (2017) functionalized iron oxide magnetic nanoparticles with beta-glucosidase enzyme via immobilization, from the peel isolates of *Dioscorea opposita*, towards the recognition of beta-glucosidase inhibitors. The study showed that the beta-glucosidase inhibitor present in the peel isolate was 2,4-dimethoxy-6,7-dihydroxyphenanthrene and batatasin I. The study showed that the enzyme-immobilized magnetic nanoparticles can be combined with high-performance liquid chromatography (HPLC)-mass spectroscopy (MS), towards a rapid identification of beta-glucosidase inhibitors [62].

4. Future perspectives

As discussed in aforementioned sections, it is evident that several types of biosensors utilize enzyme-immobilized nanomaterials, due to their inherent ability to advance their sensitivity, selectivity, and detection limits [63, 64]. However, these nanomaterials, which are immobilized with enzymes, also present certain limitations [65]. The fact that these biosensors must detect the desired biomolecules for a long time, including a large amount of sample in a relatively small duration, may lead to heat generation [66]. The generated heat may affect the enzyme stability and integration, which in turn affect the proficiency of the biosensors [67]. Accordingly, enzyme-devoid biosensors are gaining applicational significance in recent times towards biomedical applications [68]. However, the efficiency corresponding to such enzyme-devoid biosensors, is not on par with the enzyme-immobilized, nanomaterial-based biosensors [69, 70]. Hence, heat-resistant nanomaterials, as nanosized composites, are used recently for protecting the enzymes from the heat, as well as to improve their biosensing efficiency [71]. In the future, the addition of nanocomposites, and thermal resistive DNA nanomaterial combined with enzymes, can be beneficial for improving the efficiency of biosensors [72]. In addition, these nanocomposites will be used to immobilize distinct types of a diverse range of enzymes, towards the detection of multiple biomolecules via biosensors, with a high sensitivity [73].

Conclusions

The current chapter lays an outline corresponding to biosensors that are fabricated with nanomaterials as a transducer, where enzymes as bioreceptors are immobilized on their surface. In addition, the biosensing mechanism of the enzyme immobilized nanomaterials, their efficiency, detection limit, and sensitivity, were also discussed. Even though enzyme immobilized nanomaterials are widely beneficial biosensor fabrication, limitations such as excess heat release during detection may disintegrate the enzyme and affect the efficiency of the biosensor in the long run. Thus, alternative methods, such as enzyme-free biosensors are introduced to overcome those limitations. However, reduction of heat during the process via additional components may improve the efficiency of enzyme-immobilized nanomaterial-based biosensors, compared to enzyme-free nanomaterial-based biosensors in the future.

References

[1] J.S. Wilson, Sensor technology handbook. 2004: Elsevier.

[2] M.A. Morales and J.M. Halpern, Guide to selecting a biorecognition element for biosensors, Bioconjugate chemistry, 29 (10) (2018) 3231-3239. https://doi.org/10.1021/acs.bioconjchem.8b00592

[3] N. Bhalla, P. Jolly, N. Formisano, and P. Estrela, Introduction to biosensors, Essays in biochemistry, 60 (1) (2016) 1-8. https://doi.org/10.1042/EBC20150001

[4] I.-H. Cho, D.H. Kim, and S. Park, Electrochemical biosensors: Perspective on functional nanomaterials for on-site analysis, Biomaterials research, 24 (1) (2020) 1-12. https://doi.org/10.1186/s40824-019-0181-y

[5] C. Chen and J. Wang, Optical biosensors: An exhaustive and comprehensive review, Analyst, 145 (5) (2020) 1605-1628. https://doi.org/10.1039/C9AN01998G

[6] S. Mao and J. Chen, Graphene-based electronic biosensors, Journal of Materials Research, 32 (15) (2017) 2954-2965. https://doi.org/10.1557/jmr.2017.129

[7] M. Pohanka, The piezoelectric biosensors: principles and applications, Int. J. Electrochem. Sci, 12 (2017) 496-506. https://doi.org/10.20964/2017.01.44

[8] K. Cali, E. Tuccori, and K.C. Persaud, Gravimetric biosensors, Methods in Enzymology, 642 (2020) 435-468. https://doi.org/10.1016/bs.mie.2020.05.010

[9] S.A. Pullano, M. Greco, D.M. Corigliano, D.P. Foti, A. Brunetti, and A.S. Fiorillo, Cell-line characterization by infrared-induced pyroelectric effect, Biosensors and Bioelectronics, 140 (2019) 111338. https://doi.org/10.1016/j.bios.2019.111338

[10] V. Nabaei, R. Chandrawati, and H. Heidari, Magnetic biosensors: Modelling and simulation, Biosensors and Bioelectronics, 103 (2018) 69-86. https://doi.org/10.1016/j.bios.2017.12.023

[11] Q. Zhang, Y. Lu, S. Li, J. Wu, and Q. Liu, 20 - Peptide-based biosensors, in Peptide Applications in Biomedicine, Biotechnology and Bioengineering, S. Koutsopoulos, Editor. 2018, Woodhead Publishing. p. 565-601. https://doi.org/10.1016/B978-0-08-100736-5.00024-7

[12] V. Naresh and N. Lee, A Review on Biosensors and Recent Development of Nanostructured Materials-Enabled Biosensors, Sensors, 21 (4) (2021) 1109. https://doi.org/10.3390/s21041109

[13] J. Jeevanandam, A. Kaliyaperumal, M. Sundararam, and M.K. Danquah, Nanomaterials as toxic gas sensors and biosensors, in Nanosensor Technologies for

Environmental Monitoring. 2020, Springer, Cham. p. 389-430.
https://doi.org/10.1007/978-3-030-45116-5_13

[14] J. Jeevanandam and M.K. Danquah, Nanosensors for better diagnosis of health, in
Nanofabrication for Smart Nanosensor Applications. 2020, Elsevier. p. 187-228.
https://doi.org/10.1016/B978-0-12-820702-4.00008-8

[15] C. Acquah, Y.W. Chan, S. Pan, L.S. Yon, C.M. Ongkudon, H. Guo, and M.K.
Danquah, Characterisation of aptamer-anchored poly(EDMA-co-GMA) monolith for
high throughput affinity binding, Scientific Reports, 9 (1) (2019) 14501.
https://doi.org/10.1038/s41598-019-50862-1

[16] C. Acquah, D. Agyei, I. Monney, S. Pan, and M.K. Danquah, Chapter 7 - Aptameric
Sensing in Food Safety, in Food Control and Biosecurity, A.M. Holban and A.M.
Grumezescu, Editors. 2018, Academic Press. p. 259-277.
https://doi.org/10.1016/B978-0-12-811445-2.00007-6

[17] B. Purohit, P.R. Vernekar, N.P. Shetti, and P. Chandra, Biosensor nanoengineering:
Design, operation, and implementation for biomolecular analysis, Sensors
International, (2020) 100040. https://doi.org/10.1016/j.sintl.2020.100040

[18] N. Wongkaew, M. Simsek, C. Griesche, and A.J. Baeumner, Functional
nanomaterials and nanostructures enhancing electrochemical biosensors and lab-on-a-
chip performances: recent progress, applications, and future perspective, Chemical
reviews, 119 (1) (2018) 120-194. https://doi.org/10.1021/acs.chemrev.8b00172

[19] D.-M. Liu and C. Dong, Recent advances in nano-carrier immobilized enzymes and
their applications, Process Biochemistry, 92 (2020) 464-475.
https://doi.org/10.1016/j.procbio.2020.02.005

[20] A. Soussou, I. Gammoudi, F. Moroté, A. Kalboussi, T. Cohen-Bouhacina, C.
Grauby-Heywang, and Z.M. Baccar, Efficient Immobilization of Tyrosinase Enzyme
on Layered Double Hydroxide Hybrid Nanomaterials for Electrochemical Detection of
Polyphenols, IEEE Sensors Journal, 17 (14) (2017) 4340-4348.
https://doi.org/10.1109/JSEN.2017.2709342

[21] A. Majouga, M. Sokolsky-Papkov, A. Kuznetsov, D. Lebedev, M. Efremova, E.
Beloglazkina, P. Rudakovskaya, M. Veselov, N. Zyk, Y. Golovin, N. Klyachko, and
A. Kabanov, Enzyme-functionalized gold-coated magnetite nanoparticles as novel
hybrid nanomaterials: Synthesis, purification and control of enzyme function by low-
frequency magnetic field, Colloids and Surfaces B: Biointerfaces, 125 (2015) 104-109.
https://doi.org/10.1016/j.colsurfb.2014.11.012

[22] Y. Lu, M. Yang, F. Qu, G. Shen, and R. Yu, Enzyme-functionalized gold nanowires for the fabrication of biosensors, Bioelectrochemistry, 71 (2) (2007) 211-216. https://doi.org/10.1016/j.bioelechem.2007.05.003

[23] D. Du, J. Wang, D. Lu, A. Dohnalkova, and Y. Lin, Multiplexed Electrochemical Immunoassay of Phosphorylated Proteins Based on Enzyme-Functionalized Gold Nanorod Labels and Electric Field-Driven Acceleration, Analytical Chemistry, 83 (17) (2011) 6580-6585. https://doi.org/10.1021/ac2009977

[24] L.P. Datta, A. Chatterjee, K. Acharya, P. De, and M. Das, Enzyme responsive nucleotide functionalized silver nanoparticles with effective antimicrobial and anticancer activity, New Journal of Chemistry, 41 (4) (2017) 1538-1548. https://doi.org/10.1039/C6NJ02955H

[25] D.R. Bagal-Kestwal and B.-H. Chiang, Platinum nanoparticle-carbon nanotubes dispersed in gum Arabic-corn flour composite-enzymes for an electrochemical sucrose sensing in commercial juice, Ionics, 25 (11) (2019) 5551-5564. https://doi.org/10.1007/s11581-019-03091-5

[26] K. Korschelt, R. Ragg, C.S. Metzger, M. Kluenker, M. Oster, B. Barton, M. Panthöfer, D. Strand, U. Kolb, and M. Mondeshki, Glycine-functionalized copper (II) hydroxide nanoparticles with high intrinsic superoxide dismutase activity, Nanoscale, 9 (11) (2017) 3952-3960. https://doi.org/10.1039/C6NR09810J

[27] D. Chávez-García, K. Juárez-Moreno, C.H. Campos, J.B. Alderete, and G.A. Hirata, Upconversion rare earth nanoparticles functionalized with folic acid for bioimaging of MCF-7 breast cancer cells, Journal of Materials Research, 33 (2) (2018) 191-200. https://doi.org/10.1557/jmr.2017.463

[28] P.M. Tiwari, K. Vig, V.A. Dennis, and S.R. Singh, Functionalized Gold Nanoparticles and Their Biomedical Applications, Nanomaterials, 1 (1) (2011) https://doi.org/10.3390/nano1010031

[29] M. Rani, U. Shanker, and A.K. Chaurasia, Catalytic potential of laccase immobilized on transition metal oxides nanomaterials: Degradation of alizarin red S dye, Journal of Environmental Chemical Engineering, 5 (3) (2017) 2730-2739. https://doi.org/10.1016/j.jece.2017.05.026

[30] J. Singh, A. Roychoudhury, M. Srivastava, P.R. Solanki, D.W. Lee, S.H. Lee, and B.D. Malhotra, A dual enzyme functionalized nanostructured thulium oxide based interface for biomedical application, Nanoscale, 6 (2) (2014) 1195-1208. https://doi.org/10.1039/C3NR05043B

[31] R. Kant and B.D. Gupta, Fiber-Optic SPR Based Acetylcholine Biosensor Using Enzyme Functionalized Ta_2O_5 Nanoflakes for Alzheimer's Disease Diagnosis, Journal of Lightwave Technology, 36 (18) (2018) 4018-4024. https://doi.org/10.1109/JLT.2018.2856924

[32] N. Verma, N. Kumar, L.S.B. Upadhyay, R. Sahu, and A. Dutt, Fabrication and Characterization of Cysteine-Functionalized Zinc Oxide Nanoparticles for Enzyme Immobilization, Analytical Letters, 50 (11) (2017) 1839-1850. https://doi.org/10.1080/00032719.2016.1245315

[33] G. Hojnik Podrepšek, Ž. Knez, and M. Leitgeb, Development of Chitosan Functionalized Magnetic Nanoparticles with Bioactive Compounds, Nanomaterials, 10 (10) (2020) https://doi.org/10.3390/nano10101913

[34] L. Wang, Y. Meng, Y. Zhang, C. Zhang, Q. Xie, and S. Yao, Photoelectrochemical aptasensing of thrombin based on multilayered gold nanoparticle/graphene-TiO2 and enzyme functionalized graphene oxide nanocomposites, Electrochimica Acta, 249 (2017) 243-252. https://doi.org/10.1016/j.electacta.2017.07.179

[35] G. Lai, H. Cheng, D. Xin, H. Zhang, and A. Yu, Amplified inhibition of the electrochemical signal of ferrocene by enzyme-functionalized graphene oxide nanoprobe for ultrasensitive immunoassay, Analytica Chimica Acta, 902 (2016) 189-195. https://doi.org/10.1016/j.aca.2015.11.014

[36] L. Zhou, T. Wang, Y. Bai, Y. Li, J. Qiu, W. Yu, and S. Zhang, Dual-amplified strategy for ultrasensitive electrochemical biosensor based on click chemistry-mediated enzyme-assisted target recycling and functionalized fullerene nanoparticles in the detection of microRNA-141, Biosensors and Bioelectronics, 150 (2020) 111964. https://doi.org/10.1016/j.bios.2019.111964

[37] S. Afreen, K. Muthoosamy, S. Manickam, and U. Hashim, Functionalized fullerene (C60) as a potential nanomediator in the fabrication of highly sensitive biosensors, Biosensors and Bioelectronics, 63 (2015) 354-364. https://doi.org/10.1016/j.bios.2014.07.044

[38] Y. Su, X. Zhou, Y. Long, and W. Li, Immobilization of horseradish peroxidase on amino-functionalized carbon dots for the sensitive detection of hydrogen peroxide, Microchimica Acta, 185 (2) (2018) 114. https://doi.org/10.1007/s00604-017-2629-x

[39] H. Gonçalves, P.A.S. Jorge, J.R.A. Fernandes, and J.C.G. Esteves da Silva, Hg(II) sensing based on functionalized carbon dots obtained by direct laser ablation, Sensors and Actuators B: Chemical, 145 (2) (2010) 702-707. https://doi.org/10.1016/j.snb.2010.01.031

[40] L. Wang, L. Wei, Y. Chen, and R. Jiang, Specific and reversible immobilization of NADH oxidase on functionalized carbon nanotubes, Journal of biotechnology, 150 (1) (2010) 57-63. https://doi.org/10.1016/j.jbiotec.2010.07.005

[41] M. Mass, L.S. Veiga, O. Garate, G. Longinotti, A. Moya, E. Ramón, R. Villa, G. Ybarra, and G. Gabriel, Fully Inkjet-Printed Biosensors Fabricated with a Highly Stable Ink Based on Carbon Nanotubes and Enzyme-Functionalized Nanoparticles, Nanomaterials, 11 (7) (2021) https://doi.org/10.3390/nano11071645

[42] H. Sharma and S. Mondal, Functionalized Graphene Oxide for Chemotherapeutic Drug Delivery and Cancer Treatment: A Promising Material in Nanomedicine, International Journal of Molecular Sciences, 21 (17) (2020) https://doi.org/10.3390/ijms21176280

[43] H. Yoon, S. Ko, and J. Jang, Field-Effect-Transistor Sensor Based on Enzyme-Functionalized Polypyrrole Nanotubes for Glucose Detection, The Journal of Physical Chemistry B, 112 (32) (2008) 9992-9997. https://doi.org/10.1021/jp800567h

[44] D. Keller, A. Beloqui, M. Martínez-Martínez, M. Ferrer, and G. Delaittre, Nitrilotriacetic Amine-Functionalized Polymeric Core-Shell Nanoparticles as Enzyme Immobilization Supports, Biomacromolecules, 18 (9) (2017) 2777-2788. https://doi.org/10.1021/acs.biomac.7b00677

[45] R.M. Bezerra, R.R.C. Monteiro, D.M.A. Neto, F.F.M. da Silva, R.C.M. de Paula, T.L.G. de Lemos, P.B.A. Fechine, M.A. Correa, F. Bohn, L.R.B. Gonçalves, and J.C.S. dos Santos, A new heterofunctional support for enzyme immobilization: PEI functionalized Fe3O4 MNPs activated with divinyl sulfone. Application in the immobilization of lipase from Thermomyces lanuginosus, Enzyme and Microbial Technology, 138 (2020) 109560. https://doi.org/10.1016/j.enzmictec.2020.109560

[46] L.Y. Jun, N.M. Mubarak, L.S. Yon, C.H. Bing, M. Khalid, P. Jagadish, and E.C. Abdullah, Immobilization of Peroxidase on Functionalized MWCNTs-Buckypaper/Polyvinyl alcohol Nanocomposite Membrane, Scientific Reports, 9 (1) (2019) 2215. https://doi.org/10.1038/s41598-019-39621-4

[47] S. Asmat, Q. Husain, and A. Azam, Lipase immobilization on facile synthesized polyaniline-coated silver-functionalized graphene oxide nanocomposites as novel biocatalysts: stability and activity insights, RSC Advances, 7 (9) (2017) 5019-5029. https://doi.org/10.1039/C6RA27926K

[48] M.B. Vineh, A.A. Saboury, A.A. Poostchi, and A. Ghasemi, Biodegradation of phenol and dyes with horseradish peroxidase covalently immobilized on functionalized RGO-SiO2 nanocomposite, International Journal of Biological

Macromolecules, 164 (2020) 4403-4414.
https://doi.org/10.1016/j.ijbiomac.2020.09.045

[49] Y. Bai, Q. Guo, J. Xiao, M. Zheng, D. Zhang, and J. Yang, An inkjet-printed smartphone-supported electrochemical biosensor system for reagentless point-of-care analyte detection, Sensors and Actuators B: Chemical, 346 (2021) 130447. https://doi.org/10.1016/j.snb.2021.130447

[50] K. Kunene, M. Sabela, S. Kanchi, and K. Bisetty, High Performance Electrochemical Biosensor for Bisphenol A Using Screen Printed Electrodes Modified with Multiwalled Carbon Nanotubes Functionalized with Silver-Doped Zinc Oxide, Waste and Biomass Valorization, 11 (3) (2020) 1085-1096. https://doi.org/10.1007/s12649-018-0505-5

[51] M.A. Akhtar, R. Batool, A. Hayat, D. Han, S. Riaz, S.U. Khan, M. Nasir, M.H. Nawaz, and L. Niu, Functionalized Graphene Oxide Bridging between Enzyme and Au-Sputtered Screen-Printed Interface for Glucose Detection, ACS Applied Nano Materials, 2 (3) (2019) 1589-1596. https://doi.org/10.1021/acsanm.9b00041

[52] I.R. Suhito, K.-M. Koo, and T.-H. Kim, Recent Advances in Electrochemical Sensors for the Detection of Biomolecules and Whole Cells, Biomedicines, 9 (1) (2021) https://doi.org/10.3390/biomedicines9010015

[53] L. Singh, R. Singh, B. Zhang, S. Cheng, B. Kumar Kaushik, and S. Kumar, LSPR based uric acid sensor using graphene oxide and gold nanoparticles functionalized tapered fiber, Optical Fiber Technology, 53 (2019) 102043. https://doi.org/10.1016/j.yofte.2019.102043

[54] G. Zhu, L. Cheng, R. Qi, M. Zhang, J. Zhao, L. Zhu, and M. Dong, A metal-organic zeolitic framework with immobilized urease for use in a tapered optical fiber urea biosensor, Microchimica Acta, 187 (1) (2019) 72. https://doi.org/10.1007/s00604-019-4026-0

[55] R. Kant and B.D. Gupta. SPR Based Optical Biosensor for Acetylcholine Utilizing Enzyme Entrapped Ta2O5 Nanoflowers Assembly Encapsulated in Chitosan and rGO Matrix. in Optical Sensors and Sensing Congress (ES, FTS, HISE, Sensors). 2019. San Jose, California: Optical Society of America. https://doi.org/10.1364/FTS.2019.JTh2A.20

[56] B. Miranda, I. Rea, P. Dardano, L. De Stefano, and C. Forestiere, Recent Advances in the Fabrication and Functionalization of Flexible Optical Biosensors: Toward Smart Life-Sciences Applications, Biosensors, 11 (4) (2021) https://doi.org/10.3390/bios11040107

[57] H.B. Yildiz, R. Freeman, R. Gill, and I. Willner, Electrochemical, Photoelectrochemical, and Piezoelectric Analysis of Tyrosinase Activity by Functionalized Nanoparticles, Analytical Chemistry, 80 (8) (2008) 2811-2816. https://doi.org/10.1021/ac702401v

[58] M. Holzinger, A. Le Goff, and S. Cosnier, Synergetic Effects of Combined Nanomaterials for Biosensing Applications, Sensors (Basel, Switzerland), 17 (5) (2017) 1010. https://doi.org/10.3390/s17051010

[59] H. Jia, P. Xu, and X. Li, Integrated Resonant Micro/Nano Gravimetric Sensors for Bio/Chemical Detection in Air and Liquid, Micromachines, 12 (6) (2021) https://doi.org/10.3390/mi12060645

[60] J. Qiu, H. Peng, and R. Liang, Ferrocene-modified Fe3O4@SiO2 magnetic nanoparticles as building blocks for construction of reagentless enzyme-based biosensors, Electrochemistry Communications, 9 (11) (2007) 2734-2738. https://doi.org/10.1016/j.elecom.2007.09.009

[61] M. Eguílaz, R. Villalonga, P. Yáñez-Sedeño, and J.M. Pingarrón, Designing Electrochemical Interfaces with Functionalized Magnetic Nanoparticles and Wrapped Carbon Nanotubes as Platforms for the Construction of High-Performance Bienzyme Biosensors, Analytical Chemistry, 83 (20) (2011) 7807-7814. https://doi.org/10.1021/ac201466m

[62] S. Zhang, D. Wu, H. Li, J. Zhu, W. Hu, M. Lu, and X. Liu, Rapid identification of α-glucosidase inhibitors from Dioscorea opposita Thunb peel extract by enzyme functionalized Fe3O4 magnetic nanoparticles coupled with HPLC-MS/MS, Food & Function, 8 (9) (2017) 3219-3227. https://doi.org/10.1039/C7FO00928C

[63] R.S.J. Alkasir, M. Ganesana, Y.-H. Won, L. Stanciu, and S. Andreescu, Enzyme functionalized nanoparticles for electrochemical biosensors: a comparative study with applications for the detection of bisphenol A, Biosensors and Bioelectronics, 26 (1) (2010) 43-49. https://doi.org/10.1016/j.bios.2010.05.001

[64] C.I. Colino, J.M. Lanao, and C. Gutiérrez-Millán, Recent advances in functionalized nanomaterials for the diagnosis and treatment of bacterial infections, Materials Science and Engineering: C, (2021) 111843. https://doi.org/10.1016/j.msec.2020.111843

[65] Y. Zhang and S. Tadigadapa, Calorimetric biosensors with integrated microfluidic channels, Biosensors and Bioelectronics, 19 (12) (2004) 1733-1743. https://doi.org/10.1016/j.bios.2004.01.009

[66] P. Bhattarai and S. Hameed, Basics of biosensors and nanobiosensors, Nanobiosensors: From Design to Applications, (2020) 1-22. https://doi.org/10.1002/9783527345137.ch1

[67] T.-F. Tseng, Y.-L. Yang, M.-C. Chuang, S.-L. Lou, M. Galik, G.-U. Flechsig, and J. Wang, Thermally stable improved first-generation glucose biosensors based on Nafion/glucose-oxidase modified heated electrodes, Electrochemistry communications, 11 (9) (2009) 1819-1822. https://doi.org/10.1016/j.elecom.2009.07.030

[68] S.A. Polshettiwar, C.D. Deshmukh, M.S. Wani, A.M. Baheti, E. Bompilwar, S. Choudhari, D. Jambhekar, and A. Tagalpallewar, Recent Trends on Biosensors in Healthcare and Pharmaceuticals: An Overview, International Journal of Pharmaceutical Investigation, 11 (2) (2021) 131-136. https://doi.org/10.5530/ijpi.2021.2.25

[69] Y. Hasebe, T. Akiyama, T. Yagisawa, and S. Uchiyama, Enzyme-less amperometric biosensor for l-ascorbate using poly-l-histidine-copper complex as an alternative biocatalyst, Talanta, 47 (5) (1998) 1139-1147. https://doi.org/10.1016/S0039-9140(98)00193-3

[70] R. Baronas, Nonlinear effects of diffusion limitations on the response and sensitivity of amperometric biosensors, Electrochimica Acta, 240 (2017) 399-407. https://doi.org/10.1016/j.electacta.2017.04.075

[71] T. Adhikary, A. Nanda, K. Thangapandi, S. Roy, and S.K. Jana, Trends in Biosensors and Role of Enzymes as Their Sensing Element for Healthcare Applications, in Microbial Fermentation and Enzyme Technology. 2020, CRC Press. p. 147-164. https://doi.org/10.1201/9780429061257-10

[72] R. Antiochia, Developments in biosensors for CoV detection and future trends, Biosensors and Bioelectronics, 173 (2021) 112777. https://doi.org/10.1016/j.bios.2020.112777

[73] L. He, Y. Yang, J. Kim, L. Yao, X. Dong, T. Li, and Y. Piao, Multi-layered enzyme coating on highly conductive magnetic biochar nanoparticles for bisphenol A sensing in water, Chemical Engineering Journal, 384 (2020) 123276. https://doi.org/10.1016/j.cej.2019.123276

Nanomaterial-Supported Enzymes
Materials Research Foundations **126** (2022) 162-191

Materials Research Forum LLC
https://doi.org/10.21741/9781644901977-6

Chapter 6

Applications of Nanoparticles-based Enzymes in the Diagnosis of Diseases

Ali Haider[1†], Aqsa Kanwal [1†], Habibullah Nadeem[1], Farrukh Azeem[1], Roshan Zameer[1],
Muhammad Umar Rafique[1] and Ijaz Rasul[1*]

1. Department of Bioinformatics and Biotechnology, Government College University Faisalabad,
Pakistan

*Correspondence: ijazrasul@gcuf.edu.pk

† Contributed equally as first co-author

Abstract

Nanozymes (NSEs), which are efficient nanomaterials with enzyme-like appearances, have proved themselves highly-stable compared to the natural enzymes. They are also organized with the exclusive fundamental properties of nanomaterials such as luminescence and magnetism. Thus, in the biomedical field, their expansions demonstrate that their catalytic movements have opened up new applications as well as opportunities. Nanozymes are excellent in the informal mass production as well as long term storage. They are also helpful in the field of biomedical technology for the treatment of many diseases. They may incorporate various therapeutically effects in the anti-inflammatory, cytoprotecting, brain diseases and dental biofilms as well as in cardiovascular diseases. They have also performed as impartial therapeutics with other therapeutic approaches to increase antitumor effects. This chapter describes their fascinating applications in therapeutics the associated mechanism.

Keywords

Nanozymes (NSEs), Therapeutics, Cardiovascular Diseases, Neural Diseases

Contents

1.1   Nanomaterials

Materials or particles having the size of 1-100 nm are known as the nanomaterials as defined by the European Commission. Their criteria for classification of nanomaterials also include that size of half of the particles in a given sample should be less than 100 nm [1]. Nanomaterials or nanoparticles may be found in nature as small particle due to environmental impacts. Moreover, they are artificially manufactured via combustion or engineering, as chemical substances for the applications in various fields at very small scale. They have similar characteristics as basic chemical substances from which they are extracted. Their features including chemical conductivity, chemical sensitivity and strength, are enhanced after processing into smaller particles. They may exhibit unique physical and chemical properties as they are extracted from bulk into nano size particles and applied for different outcomes. However, they occur in nature accidently as processed due to volcanos, forest fires and heavy combustion processes as ash and its byproducts. Other processes like engine oils and welding also releases the nanomaterials as by products and due to their heterogeneous nature they are called as ultrafine particles [2]. In industries these nanomaterials are manufactured due to their wide applications evident by their unique physicochemical properties.

1.2   Enzymes

Enzymes are the biological catalysts, found in the living organisms to speed up the metabolic reactions [3]. They control different physiological mechanisms in their body. They speed up the pace of reactions, without being utilized into it, and do not alter their structure in whole processes. Without enzymes, the metabolic reactions cannot be progressed and they are vital to continue the life processes in living organisms. The catalysis by enzymes is crucial to carryout chemical and biological reactions to control the cell metabolism [4]. For examples some of those mechanisms are digestion, transformation of food particles into energy gradients and building block formation for living body.

1.3   Nanomaterials supported enzymes (NSEs)

Due to the advancements of biotechnology, the nanomaterials are being integrated with enzymes for enhancing its efficiency and immobilization, by 1916 [5]. The molecular advancements and discovery of nanotechnology enabled the researcher to integrate different enzymes with nanomaterial to improve their capabilities which are also known as nano-biocatalysts. They have wide applications in various biological processes achieved through high surface area for adsorption, encapsulations, and high relatability to form covalent bonds [6] . That was done basically due to their unique physio chemical properties e.g., specific geometric structures, reductions in mass limitations, easy transportations as well as large surface area for modified catalysis. Some other characteristics might be their stable nanostructures, property of reusability and high reactiveness. Due to stable nanostructures and highly improved performance, nanomaterials supported enzymes are applied in the bio-catalysis, water treatment, bioremediations, bio-sensing and immobilizations processes (Figure 1) [7]. The research is on the way to find further applications in different areas to improve speed, quality and viability of process through their unique properties.

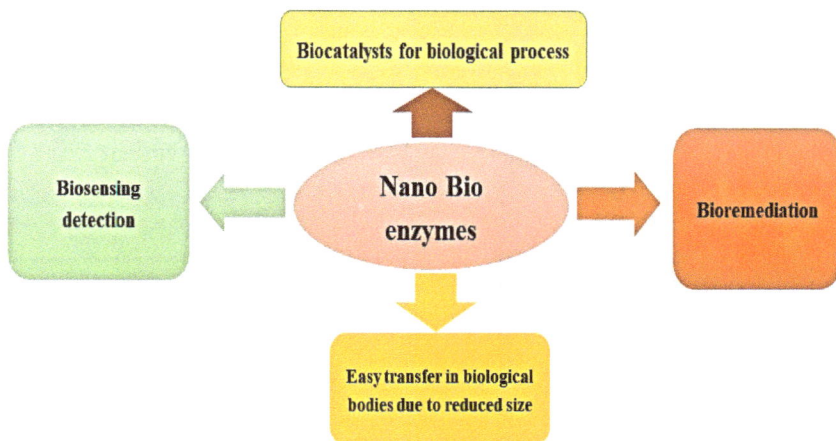

Figure 1. Applications of nanomaterial supported enzymes (NSEs)

2. Applications of nanomaterial supported enzymes (NSEs)

2.1 Role of NSEs in disease diagnosis and therapeutics

As described earlier, nanostructures have been integrated with enzymes through nanotechnology due to its large-scale applications in biomedical fields. Their unique physical and chemical properties enabled their applications in biochemical processes due to high-speed catalysis. Compared to size of nanomaterials, its structure possesses large variety of functionalities. The biological cells and tissues integrate these structures to penetrate into extracellular matrixes where that participate in different biochemical reactions. The biomedical applications of nanostructures include biofilm resistance, atherosclerotic plaques and treatment of tumors [8].

2.2   Use of NSEs in therapeutic

Due to the requirement of high-speed catalytic biological processes in case of metabolic disorders, nanostructures are applied because they have wider and unique properties. Major examples of NSEs are cerium oxides due to their efficient catalytic activity as the replacement of defected protein enzymes in patients. The anti-oxidation effects of NSEs enhance their catalytic activity for the *in vitro* conditions [9]. That establishes exploitation of the neurons and removes oxidative stress for the cardiac progenitor cells. To remove the oxidative stress in retinal degenerations, reduced ischemic cells and stent coatings, the NSEs of cerium oxide are used in their therapeutic treatments. That recently have been integrated with the phospholipid-polyethylene glycol to enhance its applications in treatment of stroke [10].

2.3   Applications of NSEs in biofilms and tumor prevention/disruption

The bacteria cells invade in the somatic tissues of living organisms with extracellular matrix and adhere to it for further pathogenic mechanisms. Those bacterial colonies are considered as biofilms that react with polymeric substances in tissue layers to cause infection [11]. These biofilms producing bacterial species are responsible for causing a wide variety of infectious diseases in humans. However, different investigations are done to explore the properties of vanadium pentoxide nanowires for the antibacterial activities and biofilm resistance. The catalytic activity and antibacterial functions induced by the vanadium pentoxide nanowires are similar to the halo peroxidases. Which are applied for catalysis of the hypobromous acid by forming the hydrogen peroxide and bromide ions [12]. The biofilm of bacterial species is degraded by NSEs due to increased levels of reactive molecules and breakdown of hydrogen peroxide. That biochemical or catalytic activity causes large scale mortalities of invaded bacteria due to degradations of polysaccharides [13].

The iron oxide, prior to biofilm resistance activities, are also applied for anti-tumor treatments as nanostructures, tumor cells are treated by nanostructures by several means (Figure 2). The synergistic effects of the iron oxide nanoparticles with the ferumoxytol are studied. That caused apoptosis in the cancerous and macrophage cells, exhibiting anti-tumor property. The macrophages cells proliferate rapidly in response to nanoparticles which lower the attacks of cancerous cells. That phenomenon is controlled by the iron oxide nanoparticles which fix the phenotype of defected genes [14].

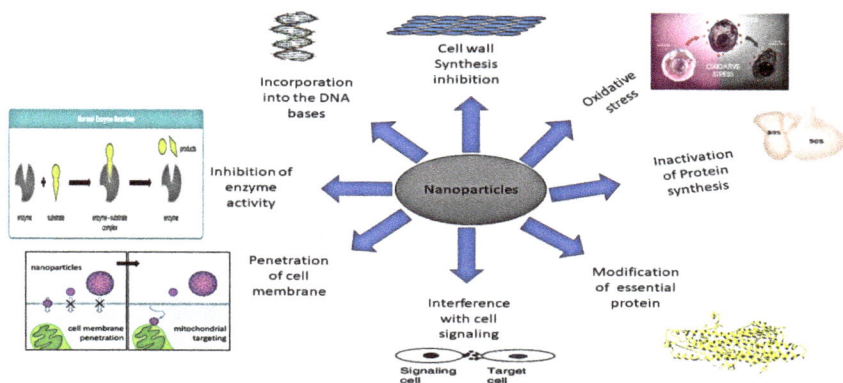

Figure 2. Illustrations of the target of tumor cells by nanostructures

2.4 The NSEs as enzymes inhibitors

Antibacterial invasion is among the main causes of death globally. The microorganisms that are narcotic are evolved as a result of the widespread utilization and misapplication of antibiotic medicines. The antibiotic resistant bacteria are providing a significant barrier to effective prevention of bacterial illness. In fact, that difficulty prompted the development of new approaches to bacterial illness management. Metallic nanoparticles (MNPs), have recently been used as antimicrobial agents, offering the ability to combat bacterial resistance to antibiotics.

2.5 Enzymatic Inhibition

The MNPs may be utilised to sense the diagnostic evaluations, regenerative medicines, radiations and thermally blistering techniques [15]. By improving pharmacokinetics and cellular uptake, the NSEs might reduce the harmful effects of antibiotics and improve their recovery efficacy when used as drug conveyers [16]. The performance of metal nano-architectures is also evaluated for antibacterial properties. That suggest they might be used to treat the contagious illnesses. Unlike conventional organic antimicrobial drugs, the MNPs have a higher carrier mobility proportion, which enhances production. That also imparts chemical, motorized, analogue, mechanical, electro-optical, magnetic and gravitational properties to the NPs that varies from own's key results. The antibacterial activity of MNPs is mostly dependent on amount, consistency and severity inside the growing media [17]. While combating bacterial infections, NPs offer different advantages compared to standard medications, including an increased accumulation of antibacterial tools within cells, avoidance of biofilm formation and fewer side effects. Antibiotic treatments are inhibited when microorganisms neutralize them until they may come into effect in the desired location. Microbes that acquired resistance to lactam antibiotics, such as penicillin, have already been widely reported. The lactam antimicrobial attacks reverse transcriptase enzymes. That enzyme is needed to for the cell wall production by hydrolyzing the aldehyde group of the four-membered polypeptide chain. The lactams are accessible in the cytoplasmic membrane of gram-negative bacteria, while -lactamase is excreted in gram-positive bacteria. The lactamase genes are contained in envelopes or transposable elements, resulting in a significant migration and transportation of specific genes to alternate microbes. Further lactamase gene alterations in such moving components might lead to multidrug resistant systems containing autoimmune reactions, sulphonamides, chloramphenicol and aminoglycosides [18].

2.6 Nanozymes for Inactivation/Inhibition of SARS-CoV-2

To battle the deadliest COVID-19 pandemic, the SARS-CoV-2 must be inactivated and inhibited. Much about SARS-CoV-2 is still undiscovered Nanozyme-mediated treatment medicines can prevent viral infections in a variety of ways, including inhibiting gene transcription and cell entrance, reducing viral incremental backups as well as inactivating the virus directly [19].

3. Role in biology and medicine

The nanoparticle exhibits multiple magnified characteristics of chemical, optical and physical in nature. They show high electrical conductivity, strong catalytic activity, heat conductance, photoelectrical properties and high reactivity. Nanomaterials have the physical electromagnetic properties to be integrated with biological processes which may be applied to imaging and diagnostic strategies [20]. Some of prominent applications of nanoparticles supported enzymes in medicine and biological mechanisms are given as bellow.

Construction of biomarkers for diagnosis of diseases, creation of fluorescent biological labels and molecules for detection of defected tissues.

- Gene therapy through nanostructures for induction in the recipient organism.

- Drug doping and delivery systems through nanostructure.

- Genetic engineering and tissues replacement.

- Diagnostic strategy for detection of disorders and metabolic diseases in organisms

- Protein and antibodies formations.

- The tumor cells apoptosis with drugs and rays.

In medicine, the drug delivery system is supported with nanoparticles for magnifying the catalytic activities. The benefits of nanostructure associated enzymes include:

- By using nanoparticles, the target drugs can be formulated.
- The nanoparticles are widely utilized by medical sciences for enhancing catalytic activities during metabolic defection.

4.   Nanozymes for sensing applications

For analytical and diagnostic processes, the nanostructures are utilized as metal oxide and non-metals oxides nanozymes in context of physiology. The detections and diagnosis of diseases are divided into two principal categories (Figure 3).

- The detection of diseases can be accompanied by integration of nanozymes and the drug agent.
- The indication of tumor cells on targeted tissues can be done with nanozymes.
- The markers and biomolecules assist in the detection of the tumor tissues.

Figure 3. Applications of the NSEs in different domains of medical fields

5.   Cancer tumor and bacterial detection

Globally the statistics of cancer patients is increasing due to various environmental factors and contaminations. That fatal disease may lead to deaths over the globe, but the diagnosis and treatment of different types of cancers is now possible due to extensive nanotechnology and synthesis of nano materials associated enzymes to compete with it. However, more

Materials Research Forum LLC
https://doi.org/10.21741/9781644901977-6

advances are needed to enhance the capabilities of nanomaterials for treatment of different metabolic disorders too. These strategies may assist in diagnosis of different types of cancers at early stages for the treatment. For the detection/diagnosis of cancer cells, nanozymes with ligands are applied which work as conjugation of aptamer. For example, the electrochemical cytosensor used for detection of MCF-7 circulating tumor cells which are caused by MUC-1 aptamer and MCF-7 cell membranes over expression. The copper oxide as nanozymes is used as signal-amplifying nanoprobe in humans for diagnosis of different types of cancers [21]. The tumor cells are detected through immunomagnetic sensors which are designed for the electrochemical detection. Another nanostructure discovered by various reports is folate-modified nanozymes for the detection of cancer cell. The electrochemical cancer cell detections are done by applying CuO/WO3 nanoparticles conjugated with folic acid (FA). The bacterial infections are related to more incidental deaths (25%) over the globe. That severity of infections is due to drug resistance and causes proliferation in bacterial populations [22].

6. Imaging, diagnostics and biomarker monitoring

NSEs are applied as the analytical and detective tools for the biomonitoring of many diseases. Ammonia levels in the human body are diagnosed by LaCoO3 (Nanostructure perovskites such as $LaMO_3$ where M= transition metal such as Mn, Co, Ni and Fe) nanoparticles. The detection and oxidation of ammonia and NO_x can be catalyzed by the NSEs, which can be monitored through light emission of laminal reaction. The peroxidase group of nanoparticles moved the hydroxyl radicals by conversion of hydrogen peroxide. However, the detection through hydroxyl groups in nanoparticles are applied in different diagnostic strategies and investigated through various reports. The nanomaterials of hydroxyl groups are integrated with luminol or 3, 3,5,5-tetramethylbenzidine (TMB) to magnify the catalysis and detection properties[23]. A variety of biological mechanisms are detected and imaged to check its accuracy by using nanoparticles. For instance, the glycolysis may be regulated using cofactors of nanoparticles. Moreover, the catalytic activity of glucose oxidase may be enhanced that is involved in conversion of hydrogen peroxide [24].

The imaging and diagnosis strategies are applied by using NSEs, known as enzyme-linked immunosorbent assays (ELISAs). That is implanted in the body of a living organism by conjugating it with anti-bodies. The iron oxide is investigated through nanotechnology as a nanoparticle and applied for the detection of hepatitis B. That interpreted the reaction between the hydrogen peroxide and TMB as ELISA assays. In that assay, the horseradish peroxidase may be replaced by the iron oxide [25]. Another example of modification of the ELISA nanozyme by integration of ferritin (M-HFn) is through encapsulation. That is used for the detection of transferrin. That is frequently used and detected at tumor tissues through iron oxide nanoparticles. The transferrin receptors are associated with tumor cells and contain ferritin-based coating [26].

Catalysts for reaction boasting for 3, 3'-diaminbenzidine (DAB) and hydrogen peroxide which formulates the DAB-polymer are brown in color. The nanoparticles may be helpful in detection of tumor cells of different cancers by brown staining. The diagnosis and detection of tumor cells can be done by targeting tumor cells through nanoparticles of fluorescent staining by using the ferritin (FITC-HFn). That technique is applied to differentiate between the tumor and normal cells due to sensitive detection of nanomaterial-based enzymes [27]. The Engineered nanoparticles (ENPs) have major applications in the domain of biomedical sciences like cancer therapy, drug delivery gene delivery, biomolecule detection, tissue engineering as well as diagnosis. By various purposes several types of ENPs have been synthesized and categorized (Figure 4). By biological activity, few inorganic ENPs (aluminum NPs (AlNPs), carbon-based NPs, copper NPs (CuNPs)) are characterized [28].

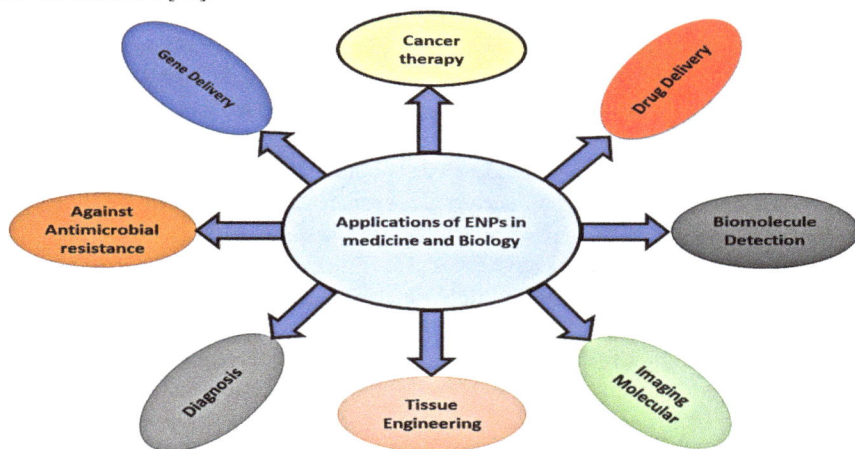

Figure 4. Applications of engineered nanoparticles (ENPs) in different fields of medicine and biology

7.   Role in HIV reactivation

The natural enzymes in living organisms can be replaced with the nanozymes having nanomaterials or particles in case of metabolic defects. The applications of NSEs for the detection and treatment of defective genes and tissue parts due to the viral and bacterial attack [29]. The laboratorial studies have revealed the applications of the nanozymes due to their unique physical and chemical properties for the artificial insertion. For instance, a nanozymes for neuroprotection and reduction of inflammation is used as ceria-based nanoparticles. The peroxides and oxidation activity of nanozymes broaden their applications as iron oxide carries the antibacterial activity to reduce biofilm production, in case of bacterial infections [30].

The ferumoxytol, (FDA) is another NSE that is used for the detection of HIV virus and inhibits tumor growth in the living organisms. The nano enzyme of vanadium pentoxide (V2O5) has same catalytic activity as glutathione peroxidase (GPX). Both exhibit the mimic behavior in biological processing [31].

The molecular technique of nanozymes of anti-oxidants in nature has applications for the diagnosis and detection of acquired immunodeficiency syndrome (AIDS) caused by human immunodeficiency virus (HIV-1). Its latency causes major issue in the diagnosis and detection. The healthy cells on the induction of virus become replication-competent and somehow, transcriptionally silent[32]. The HIV basically causes the oxidative stress by using latency resources of NF-κB and proceeds transcriptional activation through long terminal repeat (LTR). For HIV latency establishments and promotions of cellular antioxidants, the iron oxide is applied clinically which may induce long term response. The infections of HIV impose the oxidative stress as investigated through different studies and elaborates the glutathione (GSH) levels in biological tissues. The reduction in Measurement of Glutathione-redox potential (E_{GSH}) levels cause the decline in viral latency and increases in E_{GSH} Levels reactivate the HIV-1 reactivation. In case of HIV, the oxidative stresses are necessary to remove for decline in HIV viral latency and nanozymes are best applied techniques for anti-oxidative activity to deactivate the virus [33].

8.   Nanozymes for live cell and organelle imaging

Diagnosis of the clinical illness largely depends on cytological testing. Cerebrospinal fluid, exfoliated cells from blood, mucous liquid and pleural fluid may offer a wealth of medical data like cell proportion as well as cell types which can be utilized for screening cancer patients. The cytological smear, flow cytometry and nucleic acid testing are the most extensively utilized cytological detection procedures nowadays. Contrary to that the traditional procedures have significant technological requirements. Moreover, they are also time consuming, and expensive. The utilization of nanozyme-driven color reactions for qualitative and quantitative cytological analysis has expanded their use for circulating tumor cells (CTCs) detection [34]. By accelerating the oxidation process of the substrate TMB (Tetramethylbenzidine) into blue-colored products during colorimetric technique, the magnetic extraction and imaging of CTCs can be achieved at the same time via targeted Fe_3O_4 nanozymes conjugated with the specific antibody [35]. UV-vis (ultraviolet–visible) measurements can also be used to quantify the CTCs that have been visualized. The nanozymes successfully identify thirteen melanoma CTCs/mL blood in about fifty minutes, and for the TMB colorimetric development around 0.2mg/ml Fe_3O_4 is used. Later, an ultrasensitive electrochemical CTCs detection technique based on Fe_3O_4 NPs is devised [36]. The proposed nanozymes cyto-sensor exhibited significant detection accuracy for the measurement of MCF-7(Michigan Cancer Foundation-7) CTCs under optimal experimental conditions [37]. In addition to identify CTCs, catalytic activity of nanozymes is used to build the real-time detection probes in living cells for organelle imaging. For instance, a heterogeneous palladium nanozyme is developed that may successfully facilitate bio-orthogonal *processes in situ* using light. That may allow exact imaging of mitochondria in living cells [38]. Moreover, heterogeneous palladium catalysts are used for reversible light-controlled biorthogonal catalysis in living cells. In addition to CTCs and organelle imaging recognition, there are various nanozyme-based colorimetric approaches for a particular disease tomography, AIDS, diabetes, neurodegenerative diseases, infectious diseases and jaundice [39]. In blood serum the process of identifying free bilirubin is done by gold nanoclusters that are used as colorimetric probe. A nanozyme based method, compared to established methods (for instance Polymerase Chain Reaction, ELISA, and cell flow cytometry), offers a wider range of possibilities for organelle imaging. In fact they are faster, cost-effective, and easier to use [40].

In addition to diagnosis of cancer, the nanozymes have been widely used for imaging many other diseases including inflammation, infections and several neurological disorders (Table 1). That, my allow imaging exposure of biofilm-associated infections caused by diverse or mixed bacteria species [41]. For multimodal imagination of DNA base-excision improve in living cells, MnO_2 NSEs are created as intracellular catalytic DNA circuit producers. The MnO_2 nano sheet is used as a cofactor source for DNA nanozymes[42].

Table 1. Nanozymes for disease imaging

	Nanomaterials	Physical characteristics	Enzyme-mimicking activity	Bio-markers	Disease
Metal Oxide	Fe$_3$O$_4$	Magnetism	POD	CSPG	Melanoma CTCs
	Fe$_3$O$_4$	MRI	POD	----	Tumor theranostic
	M-HFn	Targeting	POD	TfR1	Tumor tissue, high-risk plaque tissues
	M-HFn	MRI	POD	----	Visualization of breast cancer cells
	Fe$_3$O$_4$@Pt	Magnetism	POD	HER2	Point-of-care bioassay
	Fe$_3$O$_4$ /rGO	Magnetism	POD	Ach	Neuropsychiatric disorders
	Co$_3$O$_4$	----	POD	VEGFR	Tumor tissue
	Fe-PDAP NFs	MRI	CAT	----	Multimodal tumor theragnostic
	CePO$_4$: Tb, Gd nanospheres	MRI and fluorescent imaging	POD	----	Multimodal imaging
	MnO	MRI	SOD	----	Tumor theranostic
	MnO$_2$	DNA zyme cofactor supplier	----	----	Living cell BER pathway
	MnO$_2$ nanosheet	----	CAT	----	UCL/PDT/RT imaging
	CuO	CL	POD	CEA	Tumor diagnosis
	Gd (OH)$_3$ and Gd$_2$O$_3$ nanorods	MRI	POD	L-cysteine	Cardiovascular and neurotoxic disease
	Prussian blue	MRI	CAT	H$_2$O$_2$	Ultrasound imaging
Noble metal	Ag	SERS	POD	CRP	Inflammatory
	Ag	Dark-field imaging	POD	HER2	Quantitative analysis of cancerous tissue
	Au	Localized SPR	Gox	ATP	Real-time imaging of targets
	Au	Two-photon photo-luminescence	POD	Integrin GPIIb/IIIa	Quantification of membrane proteins on the cell surface

	PtCo	Magnetism	OXD	----	Cancer cell imaging
	Au nanoclusters	Photo stimulated enzyme mimetics	OXD	Trypsin	Pancreatitis
	Au/Ag	----	POD	Ach	Parkinson's and Alzheimer's disease
	Pt@mSiO$_2$	----	POD	BRCA1/2	Breast cancer
	PtNPs	----	CAT	BNP, CEA.et.al	Point-of-care diagnostics
Composite Nano-Materials	Au-Fe$_3$O$_4$	Fluorescence and MRI	POD	----	Dual modal imaging cancer cells
	GO-Fe$_3$O$_4$	Magnetism	POD	Glucose	Diabetic
	FA-PtNPs/GO	----	POD	FAR	MCF-7 cancer cell imaging
	CoxFe$_{3-x}$O$_4$	Magnetism	POD	DA	Schizophrenia
	Fe$_3$O$_4$@MIL-100(Fe)	Magnetism	POD	Cholesterol	Coronary heart, myocardial infarction and stroke
	Ag@Au-Fe$_3$O$_4$	Magnetism	POD	Human IgG	Protein biomarker Detection
	ZnFe$_2$O$_4$@MWNTs	Magnetism	POD	CEA	Tumor diagnosis
	Prussian-blue/manganese dioxide	MRI, PA imaging and PT imaging	CAT		Oxygen regulation of the xenografted breast cancer
	V$_2$O$_5$-PDA-Au NPs	----	POD, Gox	Glucose	Diabetes
	NaYF4: Yb, Er		POD	Uric acid	Hyperuricemia, renal impairment and liver disease
	Fe-Co	----	POD	Glucose	Diabetes

9. The role of nanozymes in cardiovascular diseases (CVDS)

Worldwide, the cardiovascular diseases (CVDs) account for the main cause of deaths. According to the World Health Organization (WHO), the CVDs affected lives of 17.9 million individuals in 2016, accounting for 31% of all fatalities worldwide. That causes a financial load on under-developed and developing nations because a great number of people die when they are still young and active. The cardiac fibroblasts, cardiomyocytes, neural cells and vascular cells are among the cells that make up the heart [43]. CVDs are caused by any dysfunction or alteration in these endothelium, smooth, or connective cells. Ischemic heart disease, cerebrovascular illnesses, Marfan syndrome, arrhythmia, heart failure, stroke, thrombosis, vascular diseases, pericardial disease and cardiomyopathies are

Nanomaterial-Supported Enzymes   Materials Research Forum LLC
Materials Research Foundations **126** (2022) 162-191   https://doi.org/10.21741/9781644901977-6

among the ailments covered under the term CVDs. Depending on the severity and risk of CVDs, corresponding treatments vary. To minimize tissue damage and capillary rupture, all treatment strategies for CVDs focus on improving blood flow or preventing pressure from being placed on heart walls. The most popular approach for dissolving clots and restoring endothelial membrane flexibility in blood arteries is statin medication [44]. X-ray computed tomography (CT) and electrocardiography (ECG) are the most often utilized CVD diagnostic tools [45]. Biomimetic materials based on nanotechnology have recently attained a lot of attention for scaffold creation. These scaffolds are made up of nanomaterials that help with tissue repair and regeneration by providing physical, electromagnetic and mechanical support. They can help tissue function and repair by seeding cells at injury site or tissue degradation. Nano-polymeric coated biodegradable stents are now being investigated to address concerns with current stents. These new stents have the potential to improve drug release patterns while also lowering platelet adhesion rates. Anti-thrombogenic and blood compatible stents are made from nanocomposite polymers such as poly (lactic-co-glycolic acid) (PLGA) and polycaprolactone (PCL). For the medication of CVDs, nanomedicines with various properties and compositions are being studied extensively [46]. The silica-based nanoconjugates, polymeric nanoparticles, micelles, surface-modified nanostructures, niosomes, exosomes, nanofibers, nanotubes, metallic NPs, dendrimers, liposomes, hybrid nano-systems, immune modified nano-shells, and PEGylated nanospheres are its inadequate examples (Figure 5) [47].

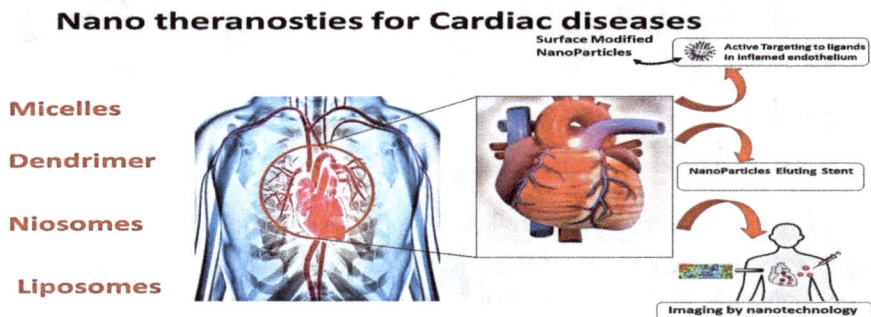

Figure 5. Applications of nanotheranostics for cardiovascular diseases (CVDs)

Table 2. Detection methods to detect nanostructures

Detection Methods	Nanostructures	Limit of Detection	References
Nano secondary ion mass spectrometry (SIMS)	Co3O4, AuNPs	7 (NanoSIMS 50 L)	[67]
Electrochemistry (EC)	Reduced graphene oxide	0.97 pg/mL	[68]
X-ray computed microtomography	CeO_2-NMs	~ 10 μm	[69]
Surface plasmon resonance (SPR)	AuNPs, Fe3O4 NPs	3.75 ng/mL	[70]
Dynamic light scattering (DLS)	Nano gold size	~0.1nm to ~ 10μm	[69]

10. Diagnosis of CVDs

If CVDs are detected early enough, then they can be treated. The diagnosis of CVDs has also been made easier due to recent advances in nanotechnology (Figure 6) [48]. Nanoscale contrast agents have evolved as multifunctional entities capable of diagnosing heart problems early on (Table 2). In comparison to traditional diagnostic techniques, nano-sensors have the advantage of being able to combine a variety of imaging agents while also being able to load medications for active targeting [49]. Due of the larger surface areas, they enable for the chelation of multivalent targeting moieties. Light scattering, fluorescent, electron-dense, paramagnetic, multimodal, or radioactive particles, generally known as nano-sensors, are used to enhance contrast in diagnostic and imaging nanostructures.

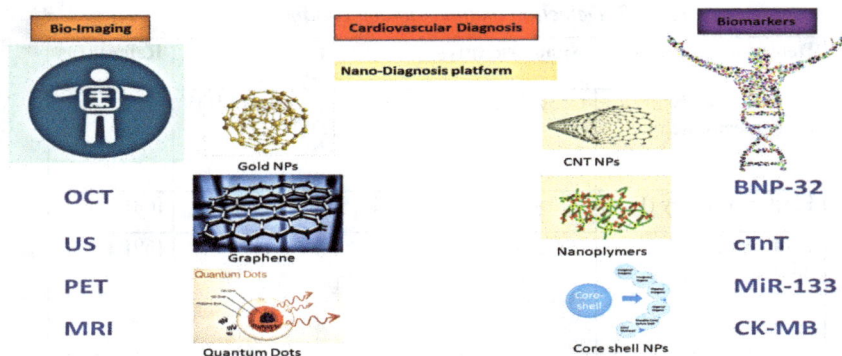

Figure 6. Diagnosis methods for the CVDs

11. Applications of Nanozymes in the treatment of CVDs

Nanozymes show a critical role while treating CVDs. The uses of Nanomaterials are directly linked to medicinal research. They can be used to provide strong alternatives to CVDs [50]. Many novel tactics may be able to improve the effectiveness of these medicines. In the last few decades, advances have been achieved for the treatment of CVDs using nanomaterials. NPs are thought to be safe and effective platforms for a variety of medicines with limited therapeutic utility due to toxicity or undesirable pharmacokinetic features (Table 3) [51].

12   The role of nanozymes in cyto-protecting

Platinum is employed as a catalyst in a wide variety of industrial processes. It is studied that the Pt (Platinum) NPs act as both SOD (Superoxide dismutase) mimics and scavengers of H_2O_2 CAT, and O_2 [52]. Additionally, conventional approaches utilized poly (acrylic acid) (PAA) in the research procedure to increase the biocompatibility of the materials. To fabricate Pt nanostructures in situ, the nano carrier apoferritin protein shell is utilized. These ferritin-platinum nanoparticles are non-toxic, stable and exhibit remarkable catalase-like activity. Previously it is suggested that Pt NPs are non-cytotoxic. Moreover, it is demonstrated that 10-4-103 ng/cm^2 (7.8610-7 -7.86 mg/l) of Pt NPs with a diameter of 20-100 nm has no effect on cellular metabolism or mortality [53]. Additionally, hybrid nanomaterials have been extensively invented. Due to their strong conductivity, absorption capabilities, vast surface area and exceptional biocompatibility, graphene oxide nanoparticles are widely used in research [54]. However, the hybrid GO-Se (Graphene oxide-selenium) nanoparticles have antioxidant capabilities comparable to those of GPx (glutathione peroxidase) for cyto-protecting. The GPx catalyzes the degradation of H_2O_2 to a non-hazardous substance in the existence of GSH (glutathione). These nanoparticles stimulate the breakdown of H_2O_2 into a less reactive compound. When Se (selenium) NPs and GO-Se nanoparticles are compared, the former demonstrated more GPx mimic catalytic efficacy. Additionally, GO-Se nanozymes are known for their great capacity for scavenging reactive oxygen species [55]. An enzyme is created using the self-assembly of polymers and protein, that nanocomposite is utilized to protect cells[56]. SP1 (Stale Protein one) is a stress-responsive protein with no resemblance to other stress proteins in its sequence. The electro-optic interaction between the polymer and protein can self-assemble and form nanowires with properties like SOD and GPx (Figure 7). These nanocomposites operate as a scavenging agent for excess reactive oxygen species and help to maintain intracellular homeostasis [57].

Table 3. Characteristics of different Nanocarriers

Types of Nanocarriers	Drugs used in the Treatment	Biological Function	References
Liposome			
Doxorubicin encapsulated liposomes	Doxorubicin (DOX)	Improve the accumulation of DOX in the tumor	[71]
Liposomal nanoparticles coated with polyethylene glycol	Prednisolone phosphate	Ideal for atherosclerotic disease	[72]
X-ray triggerable	verteporfin	Produce singlet oxygen	[73, 74]
Polymer Nanoparticles			
Dendrimer	Sulfamethoxazole	Strep throat and flu	[75]
Miceller	porphyrin	Diagnosis and Bioimaging	[76]

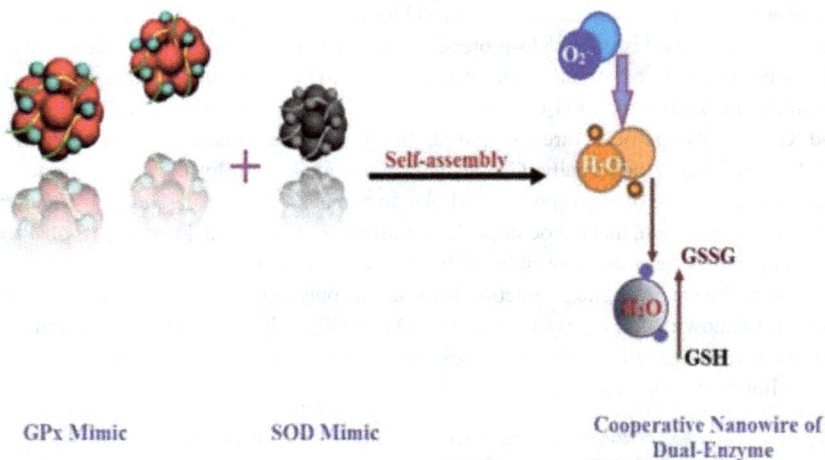

Figure 7. Self-assembly of the GPx and SOD to form a nanocomposite

13. Advances of nanozymes in the neural disorders

Throughout the world almost a billion people are affected by neurological diseases irrespective of sex, age and income. Neurological disorders like stroke and Parkinson's disease (PD) have produced huge pressure on human health. Nowadays, nanomaterials are examined to have a huge capability in the treatment of stroke because there are many deaths occur around the world due to stroke [58].

The therapeutics effects are produced by specific anatomical location of brain as well as by pathogenic mechanism because they are not understandable yet. The blood-brain barrier (BBB) is apparently considered as the main barriers. Its role is to preventing the most drugs from easy movement in the brain. Nanomaterials which are specially designed can recover the thrombolytic process of ischemic stroke [59]. Many patients and their families are damaged by inestimable due to the neurological diseases (Table 4) [60]. Through the authority of nanotechnology in 1990s, the technical barriers are vanished [61]. With the passage of time nanomaterials regularly involve in the field of visualization. Based on the properties of nanomaterials the research path of nanomaterials for neurodegenerative disorders is largely absorbed on drug delivery, early precise diagnosis and treatment of diseases. For instance, polymeric as well as lipid-based nanomaterials displayed great biocompatibility. They have potential to enter in the blood-brain barrier (BBB), with controlled release of drug. The inorganic nanomaterials are easily transformed and recognizable via imaging as well as lab methods. It is stated in 2017 that the BBB absorbency of gold nanoparticles is thoroughly associated to their size and shape. For instance with 20 nm spherical particles, the toughest penetration can be achieved [62]. In fact that make it possible to display dopamine levels in the brain as well as the gold nanoparticles may be improved by dopamine [63].

Table 4. Applications of nanomaterials for the neurodegenerative diseases

Type	Nanomaterials	Drugs Delivered	Findings	References
Lipid-based nanomaterials	Prp CsiRNA-RVG-9r-liposomes	Prp CsiRNA RVG-9rPrp Cs	Increase BBB passage rate and Development of delivery efficiency.	[77]
	Fus-liposomes-rhFGF20	LIP-FGF20	Enhance drug life, increase BBB diffusion ability, high biocompatibility, high encapsulation rate.	[78]
	RVG29-liposomes	N-3,4-Bis(pivaloyloxy)-dopamine	Higher uptake efficiency in dopaminergic cells, drug sustained release.	[79]
	PEG-liposomes-MB	GDNF+Nurr1 and neurturin	Ultrasound-guided ability, and increase drug half-life.	[78]

14. Future prospects of NSEs

Contrary to immobilization on bigger substances or immobilized enzymes, immobilizing lipids onto a range of nanomaterials offer numerous advantages [64]. In a packed-bed reactor size, use of the nanomaterials in enzyme immobilization leads to an increase enzyme packing, numerous recycle and insulation from denatured proteins of enzymes. Various nanomaterials connected amino acids might be employed in the production of renewables. In addition, the use of carbon nanotube/nanosheets on immobilized lipases/cellulases for feedstocks requires further investigation. The ability to co-immobilize several enzymes on these nanomaterials might make it easier to use different enzymes to hydrolyze complicated compounds. Further research is needed, though, to identify the technological constraints, such as intensive and costly methods for the production of nano -materials, toxic effects and the creation of security assessment criteria. The creation of efficient enzymatic solutions by the impalement of enzymes on functionalized nanoparticles substrates seems feasible, given latest events in molecular biotechnology and nanotechnology. Nanotechnology also has potential to have a huge influence on biodiesel study, from the refining of nanomaterials to the programming of enzymes for biofuels generation. In a filled to capacity reactor, only two investigations using nanomaterial attached enzymes for transesterification are done. That indicates that more research is needed in that field. Increased nanotech founder might be significant in the future to assist the implementation of green and expensive ethanol synthesis. That may fully use nanoparticles as new materials at the commercial scale. More research is needed on project management and magnitude too.

Most nanozymes have lower catalytic performance than bioactive molecules and some other biological catalysis. As a result, efforts must be made in the coming years to produce high-performance nanozymes. In this regard some work has been done. A small number of chemicals has been discovered that can effectively increase nanozyme catalytic properties.

Furthermore, nanozyme's peroxidase reaction kinetic characteristics are equivalent to those of normal HRP enzyme. Combining the operational assembly of many distinct nanozymes can open up new possibilities for the production of nanozymes with synergistically catalysis behavior of various materials. Another one of these breakthroughs might pave the way for singular sensing and theranostics that could be used in a variety of biosensors and therapeutic purposes. The majority of nanozymes are known to have catalytic properties due to redox activation of metal ions (Figure 8). Furthermore, by altering the center of the nanozymes and loading with few-earth components, the catalytic activity may be enhanced even more. More redox "hot-spots" for catalytic activity would be added as a result of such techniques, boosting nanozyme performance [65].

Over all, nanomaterials' potential problems are presently attracting a great deal of attention due to their potential implications on environmental pollution. Biocompatible materials might be developed using a "safe-by-design" approach to nanostructured materials production. Biocompatibility has also been demonstrated when biocompatible polymers like dextran are coated on the top of nanoparticles. The US Food and Drug Administration (USDA), has authorized dextran-coated iron oxide nanomaterials (Resovist) for the therapeutic usage. As a result, additional nanozymes as biodegradable catalysts for biological devices must be created [66].

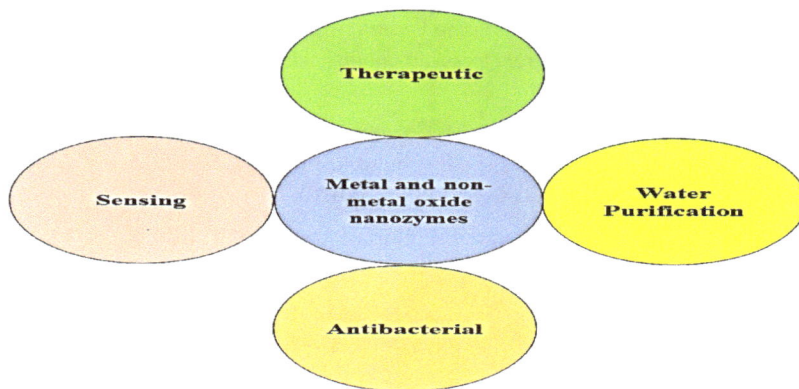

Figure 8. *Applications of metal and non-metal oxide in different biological field*

Conclusions

The biological properties of inorganic nanomaterials are revealed by the appearance of nanozymes. Nanozymes are also used in the form of natural enzymes due to their ability to report the restrictions of natural enzymes such as high cost and hard storage. During the previous era the disease diagnosis from *in vitro* to *in vivo* occurred by broadly development of nanozyme-based probes. Nanozymes have attained more attention after the detection of Fe_3O_4 nanoparticles. NSEs have many advantages like easy to prepare, good strength and easily purchasable. They are mainly useful in the cancer therapy as well as sensing etc. In this chapter, we have tried to concise the activity regulation, catalytic mechanism and present research improvement of nanozymes. Even though nanozymes also have many disadvantages compared to the natural enzymes but several exciting challenges remain. Enzymes are broadly used for different fields such as biotechnology, industry etc. Present research on applications of NSEs is still limited than natural enzymes. That shows a challenging but bright future for NSEs.

References

[1] Y. Ma, Characterization of nanomaterials in nanotoxicological analyses, Toxicology of Nanomaterials (2016). https://doi.org/10.1002/9783527689125.ch1

[2] K. Slezakova, S. Morais, M. do Carmo Pereira, Atmospheric nanoparticles and their impacts on public health, Current topics in public health, IntechOpen2013. https://doi.org/10.5772/54775

[3] R. Singhania, A. Patel, L. Thomas, M. Goswami, B. Giri, A. Pandey, Industrial enzymes, Industrial biorefineries & white biotechnology, Elsevier2015, pp. 473-497. https://doi.org/10.1016/B978-0-444-63453-5.00015-X

[4] Q.M. Dudley, A.S. Karim, M. Jewett, Cell-free metabolic engineering: biomanufacturing beyond the cell, Biotechnology journal 10 (2015) 69-82. https://doi.org/10.1002/biot.201400330

[5] R. Ahmad, M. Sardar, Enzyme immobilization: an overview on nanoparticles as immobilization matrix, J Biochemistry Analytical Biochemistry 4 (2015) 1.

[6] G. Li, P. Ma, Y. He, Y. Zhang, Y. Luo, C. Zhang, H. Fan, Enzyme-nanowire mesocrystal hybrid materials with an extremely high biocatalytic activity, Nano letters 18 (2018) 5919-5926. https://doi.org/10.1021/acs.nanolett.8b02620

[7] E.P. Cipolatti, A. Valerio, R.O. Henriques, D.E. Moritz, J.L. Ninow, D.M. Freire, E.A. Manoel, R. Fernandez-Lafuente, D. de Oliveira, Nanomaterials for biocatalyst immobilization-state of the art and future trends, RSC advances 6 (2016) 104675-104692. https://doi.org/10.1039/C6RA22047A

[8] D.S. Benoit, H. Koo, Targeted, triggered drug delivery to tumor and biofilm microenvironments, Future Medicine, 2016. https://doi.org/10.2217/nnm-2016-0014

[9] J. Wu, S. Li, H. Wei, Multifunctional nanozymes: enzyme-like catalytic activity combined with magnetism and surface plasmon resonance, Nanoscale horizons 3 (2018) 367-382. https://doi.org/10.1039/C8NH00070K

[10] T. Kang, Y.G. Kim, D. Kim, T.J.C.C.R. Hyeon, Inorganic nanoparticles with enzyme-mimetic activities for biomedical applications, 403 (2020) 213092. https://doi.org/10.1016/j.ccr.2019.213092

[11] B. Cao, X. Lyu, C. Wang, S. Lu, D. Xing, X. Hu, Rational collaborative ablation of bacterial biofilms ignited by physical cavitation and concurrent deep antibiotic release, Biomaterials 262 (2020) 120341. https://doi.org/10.1016/j.biomaterials.2020.120341

[12] J. Shi, R. Yan, Y. Zhu, X. Zhang, Determination of NH3 gas by combination of nanosized LaCoO3 converter with chemiluminescence detector, Talanta 61 (2003) 157-164. https://doi.org/10.1016/S0039-9140(03)00240-6

[13] S. Mohtashamian, S. Boddohi, Nanostructured polysaccharide-based carriers for antimicrobial peptide delivery, Journal of Pharmaceutical Investigation 47 (2017) 85-94. https://doi.org/10.1007/s40005-016-0289-1

[14] K. Fan, C. Cao, Y. Pan, D. Lu, D. Yang, J. Feng, L. Song, M. Liang, X. Yan, Magnetoferritin nanoparticles for targeting and visualizing tumour tissues, Nature nanotechnology 7 (2012) 459-464. https://doi.org/10.1038/nnano.2012.90

[15] R. Singla, A. Guliani, A. Kumari, S.K. Yadav, Metallic nanoparticles, toxicity issues and applications in medicine, Nanoscale materials in targeted drug delivery, theragnosis and tissue regeneration, Springer2016, pp. 41-80. https://doi.org/10.1007/978-981-10-0818-4_3

[16] H. Huang, W. Feng, Y. Chen, Two-dimensional biomaterials: material science, biological effect and biomedical engineering applications, Chemical Society Reviews (2021). https://doi.org/10.1039/D0CS01138J

[17] Zhang, Pornpattananangkul, C.-M. Hu, C.-M. Huang, Development of nanoparticles for antimicrobial drug delivery, Current medicinal chemistry 17 (2010) 585-594. https://doi.org/10.2174/092986710790416290

[18] J. Lin, K. Nishino, M. Roberts, M. Tolmasky, R. Aminov, L. Zhang, Mechanisms of antibiotic resistance, Frontiers in microbiology 6 (2015) 34. https://doi.org/10.3389/fmicb.2015.00034

[19] D. Liu, C. Ju, C. Han, R. Shi, X. Chen, D. Duan, J. Yan, X. Yan, Nanozyme chemiluminescence paper test for rapid and sensitive detection of SARS-CoV-2 antigen, Biosensors Bioelectronics 173 (2021) 112817. https://doi.org/10.1016/j.bios.2020.112817

[20] C. Fang, M. Zhang, Nanoparticle-based theragnostics: Integrating diagnostic and therapeutic potentials in nanomedicine, Journal of controlled release: official journal of the Controlled Release Society 146 (2010) 2. https://doi.org/10.1016/j.jconrel.2010.05.013

[21] N. Alizadeh, A. Salimi, R. Hallaj, F. Fathi, F. Soleimani, Ni-hemin metal-organic framework with highly efficient peroxidase catalytic activity: toward colorimetric cancer cell detection and targeted therapeutics, Journal of nanobiotechnology 16 (2018) 1-14. https://doi.org/10.1186/s12951-018-0421-7

[22] N. Cheng, Y. Song, M.M. Zeinhom, Y.-C. Chang, L. Sheng, H. Li, D. Du, L. Li, M.-J. Zhu, Y. Luo, Nanozyme-mediated dual immunoassay integrated with smartphone

for use in simultaneous detection of pathogens, ACS applied materials interfaces 9 (2017) 40671-40680. https://doi.org/10.1021/acsami.7b12734

[23] X.J. Yang, R.S. Li, C.M. Li, Y.F. Li, C.Z. Huang, Cobalt oxyhydroxide nanoflakes with oxidase-mimicking activity induced chemiluminescence of luminol for glutathione detection, Talanta 215 (2020) 120928. https://doi.org/10.1016/j.talanta.2020.120928

[24] M. Lobatto, V. Fuster, Z. Fayad, W. Mulder, Perspectives and opportunities for nanomedicine in the management of atherosclerosis, Nature Reviews Drug Discovery 10 (2011) 835-852. https://doi.org/10.1038/nrd3578

[25] S.E. Son, P. Gupta, W. Hur, H. Choi, H.B. Lee, Y. Park, G.H. Seong, Determination of glycated albumin using a Prussian blue nanozyme-based boronate affinity sandwich assay, Analytica Chimica Acta 1134 (2020) 41-49. https://doi.org/10.1016/j.aca.2020.08.015

[26] C. Cao, X. Wang, Y. Cai, L. Sun, L. Tian, H. Wu, X. He, H. Lei, W. Liu, G. Chen, Targeted in vivo imaging of microscopic tumors with ferritin-based nanoprobes across biological barriers, Advanced materials 26 (2014) 2566-2571. https://doi.org/10.1002/adma.201304544

[27] T.M. Allen, P. Cullis, Liposomal drug delivery systems: from concept to clinical applications, Advanced drug delivery reviews 65 (2013) 36-48. https://doi.org/10.1016/j.addr.2012.09.037

[28] G.R. Rudramurthy, M.K. Swamy, Potential applications of engineered nanoparticles in medicine and biology: An update, JBIC Journal of Biological Inorganic Chemistry 23 (2018) 1185-1204. https://doi.org/10.1007/s00775-018-1600-6

[29] R. De La Rica, D. Aili, M. Stevens, Enzyme-responsive nanoparticles for drug release and diagnostics, Advanced drug delivery reviews 64 (2012) 967-978. https://doi.org/10.1016/j.addr.2012.01.002

[30] Y. Huang, J. Ren, X. Qu, Nanozymes: classification, catalytic mechanisms, activity regulation, and applications, Chemical reviews 119 (2019) 4357-4412. https://doi.org/10.1021/acs.chemrev.8b00672

[31] A. Adhikari, S. Mondal, S. Darbar, S.K. Pal, Role of nanomedicine in redox mediated healing at molecular level, J Biomolecular concepts 10 (2019) 160-174. https://doi.org/10.1515/bmc-2019-0019

[32] L.B. Cohn, Single Cell Analysis of the HIV-1 Latent Reservoir, (2018).

[33] A. Bhaskar, M. Munshi, S.Z. Khan, S. Fatima, R. Arya, S. Jameel, A. Singh, Measuring glutathione redox potential of HIV-1-infected macrophages, Journal of Biological Chemistry 290 (2015) 1020-1038. https://doi.org/10.1074/jbc.M114.588913

[34] J. Li, J. Wang, Y. Wang, M. Trau, Simple and rapid colorimetric detection of melanoma circulating tumor cells using bifunctional magnetic nanoparticles, Analyst 142 (2017) 4788-4793. https://doi.org/10.1039/C7AN01102D

[35] F. Wang, Y. Zhang, Z. Du, J. Ren, X. Qu, Designed heterogeneous palladium catalysts for reversible light-controlled bioorthogonal catalysis in living cells, Nature communications 9 (2018) 1-8. https://doi.org/10.1038/s41467-017-02088-w

[36] L. Tian, J. Qi, X. Ma, X. Wang, C. Yao, W. Song, Y. Wang, A facile DNA strand displacement reaction sensing strategy of electrochemical biosensor based on N-carboxymethyl chitosan/molybdenum carbide nanocomposite for microRNA-21 detection, Biosensors Bioelectronics 122 (2018) 43-50. https://doi.org/10.1016/j.bios.2018.09.037

[37] M. Li, Y.-H. Lao, R. Mintz, Z. Chen, D. Shao, H. Hu, H.-X. Wang, Y. Tao, K. Leong, A multifunctional mesoporous silica-gold nanocluster hybrid platform for selective breast cancer cell detection using a catalytic amplification-based colorimetric assay, Nanoscale 11 (2019) 2631-2636. https://doi.org/10.1039/C8NR08337A

[38] F. Wang, Y. Zhang, Z. Du, J. Ren, X. Qu, Designed heterogeneous palladium catalysts for reversible light-controlled bioorthogonal catalysis in living cells, Nature communications 9 (2018) 1-8. https://doi.org/10.1038/s41467-017-02088-w

[39] M. Santhosh, S. Chinnadayyala, A. Kakoti, P. Goswami, Selective and sensitive detection of free bilirubin in blood serum using human serum albumin stabilized gold nanoclusters as fluorometric and colorimetric probe, Biosensors Bioelectronics 59 (2014) 370-376. https://doi.org/10.1016/j.bios.2014.04.003

[40] A.R. Collins, B. Annangi, L. Rubio, R. Marcos, M. Dorn, C. Merker, I. Estrela-Lopis, M.R. Cimpan, M. Ibrahim, E. Cimpan, High throughput toxicity screening and intracellular detection of nanomaterials, Wiley Interdisciplinary Reviews: Nanomedicine Nanobiotechnology 9 (2017) e1413. https://doi.org/10.1002/wnan.1413

[41] G.Y. Tonga, Y. Jeong, B. Duncan, T. Mizuhara, R. Mout, R. Das, S.T. Kim, Y.-C. Yeh, B. Yan, S. Hou, Supramolecular regulation of bioorthogonal catalysis in cells using nanoparticle-embedded transition metal catalysts, Nature chemistry 7 (2015) 597-603. https://doi.org/10.1038/nchem.2284

[42] R. Yu, R. Wang, Z. Wang, Q. Zhu, Z. Dai, Applications of DNA-nanozyme-based sensors, Analyst 146 (2021) 1127-1141.

[43] P. Zamani, N. Fereydouni, A. Butler, J.G. Navashenaq, A. Sahebkar, The therapeutic and diagnostic role of exosomes in cardiovascular diseases, Trends in cardiovascular medicine 29 (2019) 313-323. https://doi.org/10.1016/j.tcm.2018.10.010

[44] J.H. Park, D. Dehaini, J. Zhou, M. Holay, R. Fang, L. Zhang, Biomimetic nanoparticle technology for cardiovascular disease detection and treatment, Nanoscale horizons 5 (2020) 25-42. https://doi.org/10.1039/C9NH00291J

[45] C. Shi, H. Xie, Y. Ma, Z. Yang, J. Zhang, Nanoscale technologies in highly sensitive diagnosis of cardiovascular diseases, Frontiers in Bioengineering Biotechnology 8 (2020) 531. https://doi.org/10.3389/fbioe.2020.00531

[46] M. Sevostyanov, A. Baikin, K. Sergienko, L. Shatova, A. Kirsankin, I. Baymler, A. Shkirin, S. Gudkov, Biodegradable stent coatings on the basis of PLGA polymers of different molecular mass, sustaining a steady release of the thrombolityc enzyme streptokinase, Reactive Functional Polymers 150 (2020) 104550. https://doi.org/10.1016/j.reactfunctpolym.2020.104550

[47] B. Maleki, H. Alinezhad, H. Atharifar, R. Tayebee, A.V. Mofrad, One-pot synthesis of polyhydroquinolines catalyzed by ZnCl2 supported on nano Fe3O4@ SiO2, Organic Preparations Procedures International 51 (2019) 301-309. https://doi.org/10.1080/00304948.2019.1600132

[48] S.K. Metkar, K. Girigoswami, Diagnostic biosensors in medicine-a review, Biocatalysis agricultural biotechnology 17 (2019) 271-283. https://doi.org/10.1016/j.bcab.2018.11.029

[49] J. Liu, T. Lécuyer, J. Seguin, N. Mignet, D. Scherman, B. Viana, C. Richard, Imaging and therapeutic applications of persistent luminescence nanomaterials, Advanced drug delivery reviews 138 (2019) 193-210. https://doi.org/10.1016/j.addr.2018.10.015

[50] F. Sabir, M. Barani, M. Mukhtar, A. Rahdar, M. Cucchiarini, M.N. Zafar, T. Behl, S. Bungau, Nanodiagnosis and nanotreatment of cardiovascular diseases: An overview, Chemosensors 9 (2021) 67. https://doi.org/10.3390/chemosensors9040067

[51] S. Gurunathan, M.-H. Kang, M. Qasim, J.-H. Kim, Nanoparticle-mediated combination therapy: two-in-one approach for cancer, International journal of molecular sciences 19 (2018) 3264. https://doi.org/10.3390/ijms19103264

[52] J. Fan, J.-J. Yin, B. Ning, X. Wu, Y. Hu, M. Ferrari, G.J. Anderson, J. Wei, Y. Zhao, G. Nie, Direct evidence for catalase and peroxidase activities of ferritin-platinum nanoparticles, Biomaterials 32 (2011) 1611-1618. https://doi.org/10.1016/j.biomaterials.2010.11.004

[53] H. Nakanishi, T. Hamasaki, T. Kinjo, H. Yan, N. Nakamichi, S. Kabayama, K. Teruya, S. Shirahata, Low Concentration Platinum Nanoparticles Effectively Scavenge Reactive Oxygen Species in Rat Skeletal L6 Cells, Nano Biomedicine Engineering 5 (2013). https://doi.org/10.5101/nbe.v5i2.p76-85

[54] Y. Cao, Y. Ma, M. Zhang, H. Wang, X. Tu, H. Shen, J. Dai, H. Guo, Z. Zhang, Ultrasmall graphene oxide supported gold nanoparticles as adjuvants improve humoral and cellular immunity in mice, Advanced Functional Materials 24 (2014) 6963-6971. https://doi.org/10.1002/adfm.201401358

[55] Y. Huang, C. Liu, F. Pu, Z. Liu, J. Ren, X. Qu, A GO-Se nanocomposite as an antioxidant nanozyme for cytoprotection, Chemical Communications 53 (2017) 3082-3085. https://doi.org/10.1039/C7CC00045F

[56] J. Jacob, J.T. Haponiuk, S. Thomas, S. Gopi, Biopolymer based nanomaterials in drug delivery systems: A review, Materials Today Chemistry 9 (2018) 43-55. https://doi.org/10.1016/j.mtchem.2018.05.002

[57] H. Sun, L. Miao, J. Li, S. Fu, G. An, C. Si, Z. Dong, Q. Luo, S. Yu, J. Xu, Self-assembly of cricoid proteins induced by "soft nanoparticles": an approach to design multienzyme-cooperative antioxidative systems, ACS nano 9 (2015) 5461-5469. https://doi.org/10.1021/acsnano.5b01311

[58] C. Ren, Y. Yao, R. Han, Q. Huang, H. Li, B. Wang, S. Li, M. Li, Y. Mao, X. Mao, Cerebral ischemia induces angiogenesis in the peri-infarct regions via Notch1 signaling activation, Experimental neurology 304 (2018) 30-40. https://doi.org/10.1016/j.expneurol.2018.02.013

[59] K.-T. Jin, Z.-B. Lu, J.-Y. Chen, Y.-Y. Liu, H.-R. Lan, H.-Y. Dong, F. Yang, Y.-Y. Zhao, X.-Y. Chen, Recent trends in nanocarrier-based targeted chemotherapy: selective delivery of anticancer drugs for effective lung, colon, cervical, and breast cancer treatment, Journal of Nanomaterials 2020 (2020). https://doi.org/10.1155/2020/9184284

[60] D.E. Bredesen, R.V. Rao, P. Mehlen, Cell death in the nervous system, Nature 443 (2006) 796-802. https://doi.org/10.1038/nature05293

[61] A.W. Hübler, O. Osuagwu, Digital quantum batteries: Energy and information storage in nanovacuum tube arrays, Complexity 15 (2010) 48-55. https://doi.org/10.1002/cplx.20306

[62] O. Betzer, M. Shilo, R. Opochinsky, E. Barnoy, M. Motiei, E. Okun, G. Yadid, R. Popovtzer, The effect of nanoparticle size on the ability to cross the blood-brain barrier: an in vivo study, Nanomedicine 12 (2017) 1533-1546. https://doi.org/10.2217/nnm-2017-0022

[63] J.-H. An, W.A. El-Said, C.-H. Yea, T.-H. Kim, J.-W. Choi, Surface-enhanced Raman scattering of dopamine on self-assembled gold nanoparticles, Journal of nanoscience nanotechnology 11 (2011) 4424-4429. https://doi.org/10.1166/jnn.2011.3688

[64] M. Bilal, H.M. Iqbal, Chemical, physical, and biological coordination: An interplay between materials and enzymes as potential platforms for immobilization, Coordination Chemistry Reviews 388 (2019) 1-23. https://doi.org/10.1016/j.ccr.2019.02.024

[65] X. Wang, W. Cao, L. Qin, T. Lin, W. Chen, S. Lin, J. Yao, X. Zhao, M. Zhou, C. Hang, Boosting the peroxidase-like activity of nanostructured nickel by inducing its 3+ oxidation state in LaNiO3 perovskite and its application for biomedical assays, Theranostics 7 (2017) 2277. https://doi.org/10.7150/thno.19257

[66] X. Hu, F. Li, F. Xia, X. Guo, N. Wang, L. Liang, B. Yang, K. Fan, X. Yan, D. Ling, Biodegradation-mediated enzymatic activity-tunable molybdenum oxide nanourchins for tumor-specific cascade catalytic therapy, Journal of the American Chemical Society 142 (2019) 1636-1644. https://doi.org/10.1021/jacs.9b13586

[67] Y.P. Kim, H.K. Shon, S.K. Shin, T.G.J.M.s.r. Lee, Probing nanoparticles and nanoparticle-conjugated biomolecules using time-of-flight secondary ion mass spectrometry, 34 (2015) 237-247. https://doi.org/10.1002/mas.21437

[68] D.-Y. Wu, J.-F. Li, B. Ren, Z.-Q.J.C.S.R. Tian, Electrochemical surface-enhanced Raman spectroscopy of nanostructures, 37 (2008) 1025-1041. https://doi.org/10.1039/b707872m

[69] A. Yan, Z.J.N.V.i.P.S. Chen, Detection methods of nanoparticles in plant tissues, 99 (2018). https://doi.org/10.5772/intechopen.74101

[70] A. Amirjani, D.F.J.S. Haghshenas, A.B. Chemical, Ag nanostructures as the surface plasmon resonance (SPR)- based sensors: a mechanistic study with an emphasis on

heavy metallic ions detection, 273 (2018) 1768-1779.
https://doi.org/10.1016/j.snb.2018.07.089

[71] A. Gao, X.-l. Hu, M. Saeed, B.-f. Chen, Y.-p. Li, H.-j.J.A.P.S. Yu, Overview of recent advances in liposomal nanoparticle-based cancer immunotherapy, 40(9) (2019) 1129-1137. https://doi.org/10.1038/s41401-019-0281-1

[72] A.S.A. Lila, K. Nawata, T. Shimizu, T. Ishida, H.J.I.j.o.p. Kiwada, Use of polyglycerol (PG), instead of polyethylene glycol (PEG), prevents induction of the accelerated blood clearance phenomenon against long-circulating liposomes upon repeated administration, 456(1) (2013) 235-242.
https://doi.org/10.1016/j.ijpharm.2013.07.059

[73] D. Chitkara, N.J.P.r. Kumar, BSA-PLGA-based core-shell nanoparticles as carrier system for water-soluble drugs, 30(9) (2013) 2396-2409.
https://doi.org/10.1007/s11095-013-1084-6

[74] W. Deng, W. Chen, S. Clement, A. Guller, Z. Zhao, A. Engel, E.M.J.N.c. Goldys, Controlled gene and drug release from a liposomal delivery platform triggered by X-ray radiation, 9(1) (2018) 1-11. https://doi.org/10.1038/s41467-018-05118-3

[75] S. Karandikar, A. Mirani, V. Waybhase, V.B. Patravale, S. Patankar, Nanovaccines for oral delivery-formulation strategies and challenges, Nanostructures for Oral Medicine, Elsevier2017, pp. 263-293. https://doi.org/10.1016/B978-0-323-47720-8.00011-0

[76] K. Subramani, W. Ahmed, Nanoparticulate drug delivery systems for oral cancer treatment, Emerging Nanotechnologies in Dentistry, Elsevier2012, pp. 333-345.
https://doi.org/10.1016/B978-1-4557-7862-1.00019-5

[77] I. Cacciatore, M. Ciulla, E. Fornasari, L. Marinelli, A.J.E.o.o.d.d. Di Stefano, Solid lipid nanoparticles as a drug delivery system for the treatment of neurodegenerative diseases, 13(8) (2016) 1121-1131. https://doi.org/10.1080/17425247.2016.1178237

[78] K. Wang, X. Zhu, E. Yu, P. Desai, H. Wang, C.-l. Zhang, Q. Zhuge, J. Yang, J.J.J.o.N. Hu, Therapeutic Nanomaterials for Neurological Diseases and Cancer Therapy, 2020 (2020). https://doi.org/10.1155/2020/2047379

[79] M. Qu, Q. Lin, S. He, L. Wang, Y. Fu, Z. Zhang, L.J.J.o.c.r. Zhang, A brain targeting functionalized liposomes of the dopamine derivative N-3, 4-bis (pivaloyloxy)-dopamine for treatment of Parkinson's disease, 277 (2018) 173-182.
https://doi.org/10.1016/j.jconrel.2018.03.019

Materials Research Forum LLC
https://doi.org/10.21741/9781644901977-7

Chapter 7

Drug Delivery using Nano-Material based Enzymes

Fatima Mujahid[1], Sara Mahmood[1], Sumreen Hayat[1], Muhammad Saqalein[1], Bilal Aslam[1], Mohsin Khurshid[1], Saima Muzammil*[1]

[1]Department of Microbiology, Government College University, Faisalabad, Pakistan

* saimamuzammil83@gmail.com

Abstract

From the last two decades the world has progressed enormously to upgrade the wellbeing of humans by revamping the disease diagnostic and treatment. To accomplish this task, the nano-biotechnology has significantly aided in the complete transformation of disease treatment. Nanomaterials have been of great interest for better drug delivery, due to their significant catalytic activities, feasibility, and reduced production cost. Moreover, the implementation of enzyme like properties, to increase better drug delivery has gained enormous attention. Modification of the nano-scaled materials to nanozymes and enzyme-responsive nanoparticles is considered as revolutionary concept in the field of theragnostic. This chapter elaborates the diversified range of nano-material based enzymes, their synthesis methods, modification strategies, and factors influencing the catalytic activity of these enzymes. Therapeutic applications of nano-material based enzymes and their limitations have also been discussed.

Keywords

Drug Delivery, Nanozymes, Nanomedicine

Contents

1.   Introduction to Nanozymes

Nanotechnology is generating enormous developments in the field of biomedicine. It involves designing, synthesis and characterization of nanoscale materials which are used to transport therapeutic agents and diagnostic tools to specific targeted sites in a precise way [1]. In vivo solubility, stability, poor bioavailability, intestinal absorption, therapeutic efficacy, side effects, plasma variation of drugs and steady delivery to the target site are the difficulties in most of the drug delivery systems. To develop and fabricate nanostructures at submicron and nanoscales that are mainly polymeric with multiple advantages is a novel strategy to overcome these difficulties.

Nanomedicine is an emerging strategy for the applications of nanotechnological systems in therapy and disease diagnosis. Currently, nanomedicines have been well appreciated because of the reason that nanostructures could be efficiently used as drug delivery agents by attaching therapeutic drugs or by encapsulating them and transport these drugs to target sites in a controlled manner. The utilization of nanoscience techniques and knowledge in medical biology and disease stoppage and remediation is an emerging field of nanomedicine. It implicates the utilization of nanorobots, nanosensors for delivery, diagnosis and sensory purposes, and actuates materials in living cells [4].

A nanoparticles-based method has been established combining both the imaging modalities of diagnosis of cancer and treatment is a good example of it. The lipid systems such as micelles and liposomes that are FDA-approved now, is the first generation of nanoparticles-based therapy. These micelles and liposomes can carry inorganic nanoparticles like magnetic or gold nanoparticles [5]. These attributes enhance the use of inorganic nanoparticles in imaging, drug delivery and in therapeutics applications. Additionally, nanostructures reportedly help in the transportation of poorly water-soluble drugs towards their target site and prevent the drugs from being deteriorated in the gastrointestinal tract. Nanodrugs exhibit good oral bioavailability since they show typical uptake phenomenon of absorptive endocytosis [6].

Targeted drug delivery using nanomaterials especially nanozymes have tremendously helped in the cure of several diseases, as the biocompatible, chemically inert nature of NPs allow the delivery of drug at the target site. Several structural modifications are performed which allow the release of drug at the site and increases the efficacy of drug with improved absorption. The structural modification involves the attachment of different moieties, including enzymes responsive crowns, MMP-2 cleavable peptides, PEGs as hydrophilic crown, hydrophilic moieties which avoid leakage of drug and prevent from the recognition by reticuloendothelial system (RES), and expedite cellular uptake as well enhanced drug delivery.

Nanomaterial-Supported Enzymes   Materials Research Forum LLC
Materials Research Foundations **126** (2022) 192-214   https://doi.org/10.21741/9781644901977-7

Nanozymes have enzyme like properties and have great potential to replace intrinsic enzymes in many fields as disease treatment/diagnosis, chemical sensing, antibacterial agents and environmental protection [8]. Nanozymes are capable to work in environments nearer to physiological conditions and respond to a series of external stimuli. The main characteristic of nanozyme is that they have size and composition dependent catalytic activity. Tremendous progress has been observed in the establishment of these nanomaterials based artificial enzymes as shown in Figure 1.

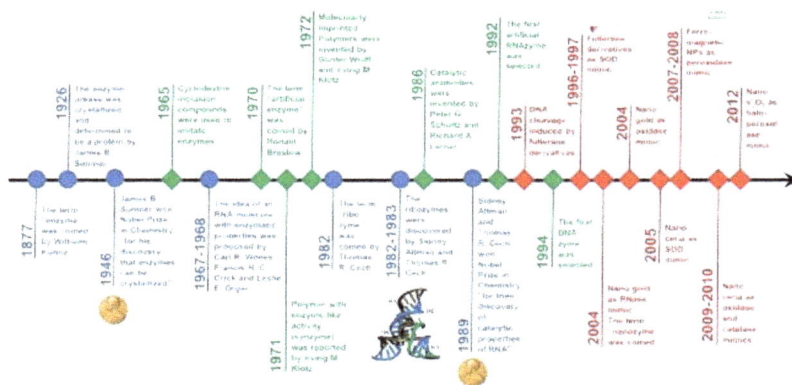

Figure 1: An outline of sequence of events in the development of nanomaterials based artificial enzymes.

Nanozymes can also overcome the limitations of natural enzymes and even the conventional artificial enzymes like high cost, low stability and difficulty in storage. By combining the controllable catalytic ability and physiochemical properties of nanomaterials, nanozymes can be used for *in vitro* spotting and *in vivo* disease monitoring and its treatment [9]. Nanozymes also provide larger surface area for bioconjugation and modifications. They have exceptional catalytic activity, fast response, multienzyme activity and self-assembly capability.

2.   Categorical distribution of nanozymes based on material type

Different types of nanomaterials are used to design nanozymes to mimic the natural enzymes. There are numerous nanomaterials that can replicate the functions of two or more enzymes. In general, metal, metal oxides materials and also carbon-based materials are used for the design of nanozymes and on the basis of which they can be arranged into three different types.

2.1   Metal-based nanozymes

Metal and metal oxide based nanomaterials have unique features such as redox chemistry, electrical and optical properties that can play an effective role in the replication of natural enzymes. For example, manganese dioxide, cerium dioxide, cobalt tetraoxide and copper oxide nanomaterials having peroxidase catalytic activity. These metal oxide nanozymes also have high surface to volume ratio and surface energy. They also exhibit low biological toxicity so that they can reside biological tissues for a long time. Metal oxide nanozymes have significant potential to replicate the activity of antioxidant enzymes such as superoxide dismutase, catalase and glutathione peroxidase. Cadmium sulfide and copper sulfide nanoparticles also exhibit similar catalytic response. Different uses of metal and metal oxides based nanoenzymes are illustrated in figure 2.

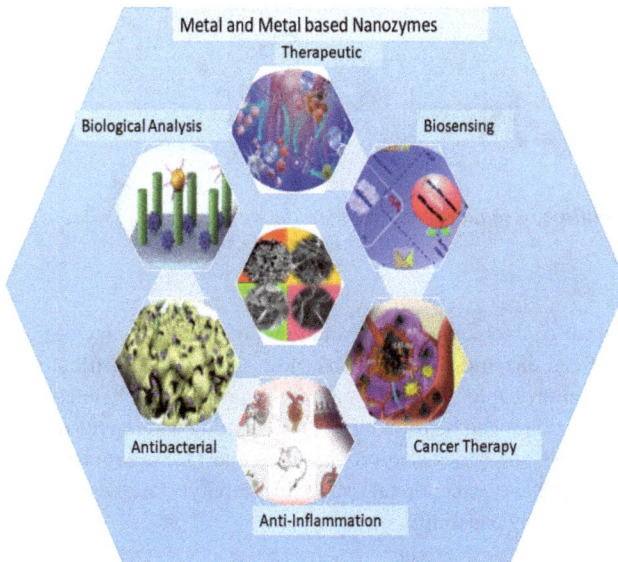

Figure 2: Different functions of metal and metal based nanozymes

Nanomaterial-Supported Enzymes   Materials Research Forum LLC
Materials Research Foundations **126** (2022) 192-214   https://doi.org/10.21741/9781644901977-7

Table 1. Some important characteristics of nanozymes and natural enzymes

Nanozymes	Regulation Mechanisms	References
Fe_3O_4	Surface modification via (polyethylene glycol ,3-aminopropyltriethoxysilane, dextran, and SiO_2)	[10]
Ag	Activator mercury (II)	[36]
CeO_2	Inhibitor (phosphate)	[37]
Pt	Inhibitor mercury (II)	[38]
VO_2	Morphology (nanorods, nanofibers and nanosheets)	[35]
Prussian Blue	pH	[39]
Ti-doped CeO_2	Composition (doping)	[40]
Graphene quantum dots	Surface modification (Benzoic Anhydride, phenyl hydrazine & 2-bromo-1-phenylethyl ketone)	[41]
Graphene oxide	Temperature, pH	[42]
Dextran-coated nanoceria	Size range of (5 nm, 12 nm, 14 nm, 100 nm)	[43]
Mn_3O_4 nanoflower	Morphology (polyhedron, nanoflowers, Cubes, flakes, hexagonal plates)	[44]
Co_3O_4 nanoplates	Morphology (nanorods, nanoplates and nanotubes)	[45]
ZrO_2 gel	pH	[46]
Au@Pt nanorods	Inhibitor (Fe^{2+}, Cu^{2+}, and NaN_3)	[46]

2.2   Fe-based nanozymes

Ferromagnetic (Fe_3O_4) nanoparticles with enzyme like properties were first reported by Gao et al. [10]. Nanozymes with Fe_2O_3 and Fe_3O_4 nanomaterials also have peroxidase catalytic activity. They used to mimic the natural peroxidases that are being utilized the oxidation of organic substrates as wastewater treatment. The size, surface modification and shape can enhance the catalytic potential of Fe-based nanozymes.

2.3   Carbon-based nanozymes

Carbon nanomaterials have well defined geometric and electronic structures that are helpful in the design of carbon based nanozymes such as graphene, carbon dots, carbon nano-spheres, graphene quantum dots, fullerenes and carbon nanotubes [11]. Carbon nanozymes provides a multifunctional platform for biomedical applications.

3.   Major Classes of nano-enzyme based on mode of action

Nanozymes can also be classified based on their mode of action.

3.1   Antioxidant nanozymes

The incomplete removal of reactive oxygen species by the antioxidant defense system of the body causes reactive oxygen stress that is the main reason of many diseases and disorders including cancer, kidney disorders, neurodegenerative diseases, diabetes, arthritis and aging. If reactive oxygen species (ROS) such as superoxide anion radicals (O_{-2}), hydrogen peroxide (H_2O_2) and hydroxyl radicals ($\cdot OH$) sustain for long time, they cause oxidative damage to lipids, DNA and proteins. They are also responsible for the activation of the apoptotic pathways in the body, which leads to severe damage. Some of the natural enzyme antioxidants and non-enzyme antioxidants are readily inactivated under non–physiological conditions and lacks the stability to fight with the reactive oxygen stress related disorders [12]. Nanozymes with antioxidant enzymes like activity have a wide range of applications in immunoassays, biosensing, therapeutics and diagnostics. Metal oxide nanozymes such as Fe_3O_4 and cerium oxide (CeO_2) are more efficient in mimicking the antioxidant enzymes of immune system like catalase, superoxide dismutase and glutathione peroxidase [13].

Prussian blue nanoparticles (PPBs) are an example of antioxidant nanozymes, which can effectively scavenge ROS with catalase activity, glutathione peroxidase activity and superoxide dismutase activity. PPBs efficiently converts harmful hydrogen peroxide (H_2O_2) into useful water (H_2O) and oxygen (O_2) and therefore avoid the damage to DNA, lipid peroxidation and protein oxidation [14]. The ultimate antioxidant nanozymes have exceptional biocompatibility; have stable antioxidant activity against oxidative damage. They can decompose reactive oxygen and nitrogen species (RONS) and avoid the inflammation caused by RONS.

3.2   Superoxide dismutase (SOD) antioxidant nanozymes

Superoxide dismutase is the first line of defense against the ROS. They are a class of metalloenzymes catalyzing the dismutation of superoxide radicals thus treating the diseases caused by oxidative stress. Superoxide dismutase (SOD) such as CeO_2 nanomaterials can decompose hydrogen peroxide (H_2O_2) through catalase like activity [15]. Cerium vanadate ($CeVO_4$) nanozymes mimic the function of superoxide dismutase (SOD), which prevents the mitochondrial damage by maintaining the superoxide level and repairing the physiological levels of Bcl-2 family proteins (the anti-apoptotic).

3.3   Pro-oxidant nanozymzes

The nanozymes which induce the oxidative stress by free radicals generation or by inhibiting the antioxidant system of the body are known as pro-oxidant nanozymes. Some metals like iron and copper and some commonly used drugs like paracetamol and anticancerous methotrexate produce excessive free radicals so regarded as pro-oxidant nanozymes [16].

Natural peroxidase enzymes are used to detoxify the free radicals; they use hydrogen peroxide (H_2O_2) to oxidize the substrate and also fight against the invading pathogens [16] Specific nanomaterials reveal catalytic activities such as peroxidase enzymes. The Fe_3O_4

nanozymes have great importance because of their antibacterial nature and peroxidase activity. Iron oxide nanoparticles have great ability to work like peroxidase enzymes. The peroxidase activity of iron oxide nanozymes is higher with smaller sized nanoparticles. Comparison studies on natural peroxidases and iron oxide nanozymes peroxidases showing that both work in controlled reaction temperature and pH while nanozymes peroxidases are more stable on higher range of temperature [16] .There are many other types of iron-based nanoparticles which have peroxidase activity like Fe-S nanosheets and iron telluride (FeTe) nanorods have greater peroxidase activity than spherical shaped nanoparticles because of large surface area [17].

4.   Nanoparticles with enzyme-responsive linker

Nanozymes are designed with the help of nanoparticles and cleavable linkers so that drugs can be attached with them and reached their specific target. These linkers have hydrophobic and hydrophilic ends through which they bind the drugs and the nanoparticles respectively. These linkers are cleavable in nature so that can be rapidly separate at the site of target. The linkers with dual characteristics are more desirable as having both binding and enzymatic activity as linker peptides also have protease activity [18]. These enzyme responsive nanoparticle linkers enhance the bioavailability of chemotherapeutic drugs at specific target and reduce the side effects. They have great potential in diagnosis and for the treatment of cancers by reducing the drawbacks of traditional chemotherapeutic drugs.

5.   Nanozymes preparation

There are numerous methods used for the preparation of nanozymes. The most common methods employed for the preparation of nanozymes are listed below.

5.1   Hydrothermal method

The hydrothermal method is applied for the synthesis of nanozymes having crystalline phase with fluctuating melting points. This process involves an aqueous reaction requiring elevated temperature and vapor pressure to re-crystallize the materials [19].Oxide based nanozymes are successfully synthesized using this method. Peroxidase-like activity displaying nanozymes and hexagonal tungsten oxide nanoflowers are synthesized by using hydrothermal method [20].

5.2   Solvothermal method

This method is one of the preliminary methods for constructing the nanomaterials. In this method a solvent with elevated temperature range (100 to 1000 °C) along with the increased pressures (from 1 to 10,000 atm) are used [21]. A wide range of nano-materials can be synthesized using this method including various semiconductors and metals, metal oxides [22]. Fe_3O_4 nanocrystals and Au/Cu (gold/copper) nanocomposites displaying oxidase-like catalytic activities are also synthesized using this method [23].

5.3 Co-precipitation method

Co-precipitation method is preferably used methods for the production of nanoparticles (NPs). This process offers the synthesis of pure and homogeneous NPs, as well as nanozymes for more than one constituent [24]. Shi et al. [25] reported the method of preparation of $CoFe_2O_4$ NPs that manifested peroxidase like activity by co-precipitation method.

6. Development of endogenous enzyme-responsive nanomaterials

Modification of conventional NPs, as stimuli-responsive NPs (SRNPs) is regarded as promising porter for the site directed drug delivery due their unique bio-responsive and physicochemical properties. These "smart" SRNPs can respond in expected manner, predominantly to external stimuli [26].

In the human body, metabolic activities direly depend on the enzymatic reactions. The idea of enzyme responsive drug release via NPs was originated from the concept of naturally enzyme catalyzed reactions. Due to their predicted and specified response to external stimuli, smart enzyme-responsive nanoparticles are applied in diversified range of anti-tumor drugs delivery [27]. Enzyme responsive NPs are considered as providential alternatives for the controlled and time specific drug release in tumor's microenvironment [28].

6.1 Synthesis of nanomaterials with enzyme-responsive core

Synthesis of nanoparticles with the enzyme responsive core involves the installation of functionally active drug entrapped in the core of NPs. The release of drug predominantly depends upon the presence and action of enzyme. Release of drug that is entrapped in the core of NPs involves confirmational changes in the structure, splintering of covalent linkages, disintegration and charge switching etc. [29].

The synthesis of enzyme responsive core NPs, implies either proteinase sensitive peptides (covalent-conjugation) to the diagnostic agent or self-assembly of peptides with natural enzyme-cleavable stretch. Matrix metalloproteinases MMPs such as MMP-2 and MMP-9 belonging to calcium-dependent zinc-containing proteinases class, are of significant importance as experimental studies have proven their metastasizing and cell invasion activities.

6.2   Nanoparticles construction with enzyme responsive crown

In order to escalate the uptake of NPs by cells, surface modifications are performed. Usually surfaces of NPs are modified with hydrophilic constituents to avoid the reorganization by reticuloendothelial system (RES), to avoid the interaction with cells, to prevent the drug-leakage. Diversified range of materials with enhanced hydrophilic properties have been considered as potent surface modifiers, including hyaluronic acid (HA), peptides, synthetic polymers, proteins etc. For the efficient discharge of encapsulated drugs, it was supposed that hydrophilic auxiliary of NPs would be slipped off as the NPs reached the target site. Enzyme-cleavable peptides are of quite significance for synthesizing NPs with enzyme responsive crown [30]. Rather than, constructing crown entirely with peptides, nanozymes can be constructed with synthetic hydrophilic polymers crosslinked with enzyme-cleavable peptides. This mechanism facilitates the triggered-drug release, by partial degradation of the polymeric-crown [28]. Li et al. deciphered the method of efficient drug release by the NPs containing synthetic polyethylene glycol (PEG) as hydrophilic crown & further modified by the adding MMP-2 responsive peptides & tumor-targeting-ligands [31]. This experiment also explained that modification of NPs with PEG was performed to increase the circulation time of drug in blood.

6.3   Modification of nanomaterials with enzyme responsive linker

Generally, cleavable linkers are necessary constituent of nanoparticles. These are used in diversified range of NPs modification such as establishing the link between the hydrophilic crown and hydrophobic core. Also, they are used in the modification of hydrophilic surfaces by ligands targeting. Ideally, cleavable linkers corroborates that the auxiliary of nanoparticles must remain joined while the drug circulation, but surely get cleaved immediately after the NPs reach the target site [18]. It is preferred in the NPs modification, if the linkers are conferred with enzyme-responsive capability. In the fabrication of enzyme-responsive linkers, peptide harboring proteases are the most commonly used. For the anti-tumor drug delivery system, MMPs are the most studied proteases and even in some cases MMPs degradable peptides have been reported as a linker. In a study, NPs with hydrophilic PEG (as cationic charge shielding surfaces) were constructed [32, 28]. The active drug loaded core was attached to PEG by MMP degradable peptide linker, as shown in figure 3. It was observed that these long nanoparticles might be localized passively in tumor-tissue. In the presence of MMP2/9 that are PLGLAG-sensitive, PEG-layer could fell off and uptake of NPs by tumor cells when, exposed to positive charge.

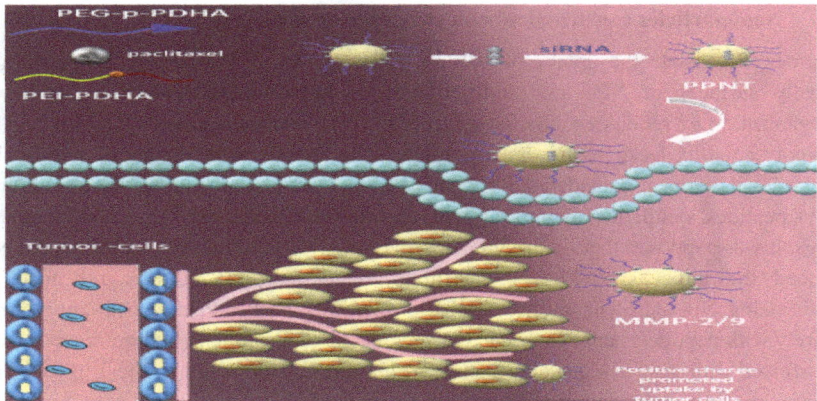

Figure 3: Diagrammatic representation of synthesis of siRNA and PPTN (paclitaxel loaded micelle) and its activation in response to MMP-2/9.

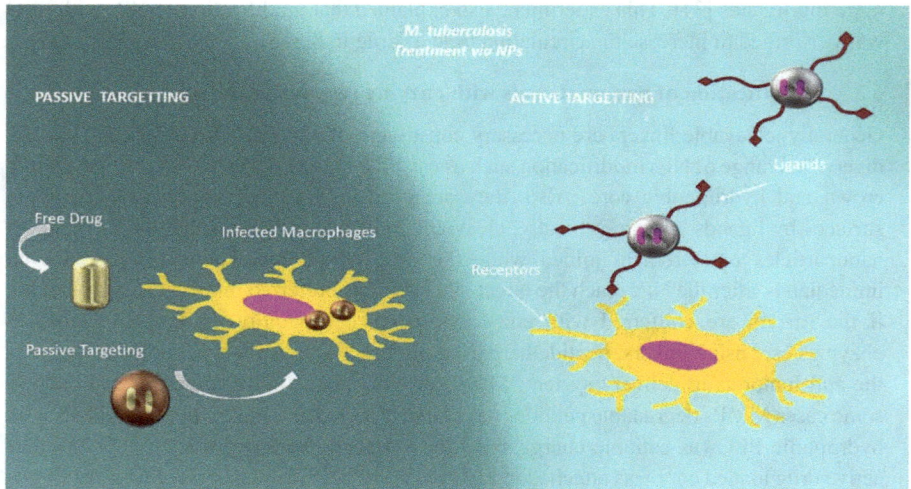

Figure 4: Active and passive targeting of NPs loaded with drugs for the treatment of TB

6.4 Nanoparticles and enzyme-responsive ligands

A cause of the elevated heterogeneity levels of tumor cells, precise target delivery of anti-tumor drugs are necessary. Several strategies have been implemented to modify the targeting-ligands. The selection and structural configuration of the target ligands entirely depend on the receptors over-expressed in the disease tissues. For the synthesis of target ligands with enzyme responsive ability, HA and peptides play a vital role [33].

7. Factors affecting nanozymes activity

The optimized functioning of nanozymes can be affected by size, shape, morphology, temperature and pH. In the following section some of the factors affecting the optimized catalytic functioning of the nanozymes will be discussed. Some of the regulation strategies of nanozymes activities have been summarized in table 2.

Nanozymes	Natural enzymes
Low cost	Increased substrate specificity
Easy for mass production	Enhanced selectivity
High catalytic efficiency	High catalytic activity
Robustness to extreme environments	Refined three-dimensional structures
Increased stability	Broad range of catalytic reactions
Long-term storage	Tunable catalytic activity
Tunable catalytic activity and types	Enhanced biocompatibility
Size- (structure-, shape-, composition-) dependent characteristics	Rational designs by computation and protein engineering
Multifunctional	
Recyclable utilization	
Efficient response to external stimuli	
Unique physiochemical properties	

7.1 Morphology

Accurate surface-structures are necessary for the efficient working of nanozymes. It was reported that enzymatic activities of $LaNiO_3$, Fe_3O_4, $CoFe_2O_4$, Co_3O_4 get significantly influenced by the morphological changes [34]. For example, increased peroxidase-like activities were represented by vanadium oxide (VO_2) nanofibers when compressed to nano sheets or nanorods due to morphological changes [35].

Nanomaterial-Supported Enzymes   Materials Research Forum LLC
Materials Research Foundations **126** (2022) 192-214   https://doi.org/10.21741/9781644901977-7

7.2   Size

In the modification of nanozymes, size of the NPs plays a vital role. Usually, smaller NPs represent increased catalytic activity as they have specific-surface area [47]. Zhang et al. [48] reported that by decreasing the size of Fe_3O_4 NPs, their peroxidase-like activity was significantly increased.

7.3   Surface modifications

Catalytic efficiency and performance of nanozymes are greatly affected by the surface modifications. Numerous methods can be used for the modification of nanozymes such as covalent immobilizations of functional groups and by electrostatic adsorption [10]. By increasing the active sites, significant increase in the catalytic activities of nanozymes are observed [49].

8.   Therapeutic applications of nanozymes

Nanoparticles have many applications in the field of biology such as magnetic nanoparticles are used in the waste water treatment, catalysis, biomedical separation and magnetic target drug delivery [45]. The activity of enzymes has been combined with the properties of nanoscaled materials to expand the applications of nanozymes. Many types of the nanozymes have been reported and their applications in the environmental remediation and biomedical technology in diagnosis, drug delivery and treatment of diseases have been reported [31]. Some of the applications of nanozymes have been mentioned below.

8.1   Cytoprotection

Nanozymes are involved in the cytoprotection as platinum (Pt) NPs serve the role of SOD mimics to scavenge the H_2O_2, CAT and $O2^{•-}$. Pt nanostructures were synthesized by using apoferritin (protein shell) as nanocarriers. These NPs were stable, non-toxic, bioactive and were having exceptional catalase like activity [28]. Similarly, graphene oxide nanomaterials have been extensively used in the research because of their large surface area, high conductivity, increased absorbing activities and good biocompatibility.

A nanocomposite having enzymatic activity was constructed by self-assembly of polymers and proteins that can be used for the cytoprotection. Protein (SP1) is a stress responsive protein, GPx, (Glutathione peroxidase) and SOD (Superoxide dismutase) catalytic center were assembled into the protein SP1 and polymer nanoparticles (PD5). These nanocomposites maintained the intracellular balance by scavenges the reactive oxygen species (ROS) [19].

8.2   Nano carriers

Nanomaterials having the potential to wrap nano-sized materials, to transport, to protect and to release at the site of action are known as nanocarriers, e.g., Liposomes and micelles are used as drug delivery systems in treating many diseases. Nanocarriers are used in therapeutics to treat different diseases such as cardiovascular diseases. Liposomes are lipids in nature having similar structure as plasma membrane so they can easily pass through the cell. These biocompatible vesicles size ranging from 80-300 nm are broadly used in nanomedicine because of their low toxicity for *in vivo* experiments. They incorporate the hydrophobic agents inward the lipid bilayer and hydrophilic therapeutic agents inside the aqueous place to protect them from degradation. Liposomes have high agent loading ability, controllable release and high stability in the biological environment. They encapsulate the drug and deliver them inside the cell at the target side (Qamar et al., 2019).

The micelles with size range of 10-100 nm have amphiphilic monolayer structure. They incorporate hydrophobic agents and deliver them on their site of action. These are stimuli-based nano-carriers for the controlled drug delivery [15].

8.3   Nanozymes as antibacterial, anti-inflammatory and antibiofilm agents

Microbial infections are becoming a major global issue, and the main response to overcome this problem is the use of antibiotics. However, the extensive use and misuse of antibiotics led to the development of multi drug resistance (MDR) bacteria, decreasing the therapeutic efficiency of antibiotics and increasing the mortality rates [49]. Moreover, most of the bacterial pathogens also have the potential to form biofilms enclosed by self-secreted extracellular polymeric substance (EPS) mainly composed of polysaccharides, proteins, lipids and extracellular DNA (eDNA). Due to the presence of EPS, biofilms forming bacteria can exhibit more resistance to antibacterial agents and can also evade the host's immune system. This situation prompts the researchers to develop novel antibacterial and antibiofilm materials effective against MDRs (Yang et al. 2015; Natsos et al., 2019). So, for nanozymes having fundamental enzyme like activities regulating the reactive oxygen species (ROS) by redox reactions have been developed. These nanozymes are more biocompatible, stable and reusable than the natural enzymes. So, the growth of different types of bacteria, gram negative, gram positive and biofilm forming, can be inhibited.

An antibacterial composite i.e., based on polyethylene glycol (PEG) functionalized molybdenum disulphide nanoflowers (PEG-MoS2 NFs) was established by Yin et al. (2016). This nano enzymes-based composite had peroxidase activity that could efficiently decompose the H_2O_2 to produce OH having accelerated anti-bacterial activity. Whereas, the generation of OH radicals made the architecture of the resistant bacteria more fragile thus wound-healing process became easier. Molybdenum disulfide in this composite material also had good photothermal activity thus providing an efficient strategy for *in vitro* rapid killing of *E. coli* (i.e., ampicillin resistant) and *Bacillus subtilis* (endospore-forming). Gold nanoparticles (AuNPs) were conjugated with graphite carbonitride (g-C$_3$N$_4$) to prepare antibacterial and antibioflms system (Wang et al. 2016). This system could effectively be used for the killing of bacteria, disinfection of wounds, and for the treatment of lung infections at biologically applicable concentrations of H$_2$O$_2$. The peroxidase activity of nanomaterial was significantly enhanced by the synergistic effect of AuNPs and g-C$_3$N$_4$ that could more efficiently decompose H$_2$O$_2$ to generate OH. It was observed that this nanomaterial-based enzyme system not only exhibited promising bactericidal activities against MDR Gram-positive and Gram-negative bacteria but also effectively decomposed pre-established multidrug resistant biofilms and prevented the formation of new ones. Oral diseases associated with biofilm production by bacteria especially *Streptococcus mutants* can be treated by nanozymes. Fe-based nanoparticles (Fe$_3$O$_4$) with CAT nanoparticles having peroxidase like activity that is useful in preventing the biofilm formation by bacteria. These CAT nanoparticles catalyze the OH formation in low pH environment that will kill the biofilm forming bacteria and also mutilate the biofilms and protect the teeth.

DNase-mimetic artificial enzyme (DMAE) was developed by AuNPs adsorption physically on the surface of colloidal, magnetic Fe3O4/SiO2 core/shell particles and then assembled a single layer of Ce(IV) on the bare AuNP (Chen et al. 2016). This system had efficient cleavage capacity for natural DNA and model substrates, having the potential to inhibit the biofilms formation in 120 hours. By the DNase activity of nanomaterials, integrity of EPS surrounding the biofilms was targeted thus increasing the bactericidal efficiency of antibiotics against biofilms forming bacteria. Cerium oxide nanoceria plays an important role in nano-therapeutics and decreases the chronic inflammation that is the part of many diseases as diabetes, arthritis, Parkinson's and heart diseases. Chronic inflammation is due to the high level of nitric oxide in the body leading to the organ damage. Nanoceria has the ability to scavenge the reactive oxygen species (ROS) and inhibit the moderators of inflammation. Cerebral ischemia/stroke is the most common cause of death in which the supply of oxygen and glucose is reduced to the brain leading to stress, blood brain barrier and ultimately the cell death. Cerium oxide nanoparticles have the ability to treat stroke. When cerium and oxygen combine in the formation of nanoparticles, they become antioxidant in nature and have the ability to treat oxidative stress [11].

8.4   Nanomaterials based targeted drug delivery to overcome tuberculosis (TB)

Due to the expeditious spread of MDR and XDR (extensively drug resistant) strains [50], TB has always been a challenge to health-sector. The poor-solubility of drug and elevated toxicities associated with the drug direly require the shifting of TB treatment towards new potent alternatives [51]. Using nanocarriers as a potent vector to administer the drug to the target site is currently applied. Currently for the treatment of TB via nanocarriers, both active and passive targeting have been implemented. NPs loaded with the drug can be administered via different routes due to heterogenous responses represented by biological system. For the treatment of TB pulmonary, oral, and intravenous routes of drug administration are preferred, but oral route is considered as the ideal one.

Delivering the NPs loaded with drug by passive targeting is advantageous as nanocarriers are ardently taken up by macrophages [52]. Due to the continuous movement of macrophages to the site of infection, drug can be proficiently transported that makes passive targeting a promising alternative for TB treatment [53]. The active targeting by NPs involves the surface modification of NPs with ligands, particular to receptors displayed by macrophages. Thus, active targeting provides enhanced treatment of TB using nano- carriers [53].

8.5   Anti-tumor drug delivery via enzyme-responsive NPs

Cancer is one of the leading cause of mortality [54] and the challenges leading to the failure of cancer treatment involves the resistance to chemo-drugs, recurrence of tumors and tumor metastasis [55]. Efflorescing advancements in the field of nanoscience has enabled us to curb the issue of increased mortality due to cancer. Advancements in the field of nanoscience have been made on the concept of dire-expression of tumor associated-enzymes to construct the enzyme responsive nanocarriers for the targeted drug release. Synthesis of enzyme-responsive NPs for targeted on-response drug release has tremendously aided in the effective cancer treatment [56].

Peroxidase-like nanozymes have taken significant attention [57]. During rapid proliferation in tumor cells, excessive amount of released reactive-oxygen species and hydroxyl radicals have been observed [58]. Peroxidase like enzymes, exclusively aid in the conversion of hydrogen-peroxide in reactive oxygen radicals, which ultimately help in the development of cancer treatment [59] that has been experimentally proved by 97% breast tumor inhibition after the treatment with peroxidase-mimetic boron NPs [60].

9.   Limitations of nanozymes

Despite the significant advantages observed by nanozymes in the wide array of diagnostic, theragnostic and cancer treatment; but still there are various challenges regarding the practical clinical applications that must be addressed. These limitations include poor selectivity of substrate; few catalytic types, poor dispersibility and nanozymes get clogged at the site with high ionic strength resulting in the reduced catalytic activity. Also, the cytotoxity associated with nanozymes are still unknown [55].

Conclusion

Over the past few years, the advancements in the horizon of nano-medicine from nanomaterials to nanozymes based drug delivery have completely revolutionized the medical field. Nanomaterials based enzymes are considered as potent alternatives for the better targeted-drug delivery due to their reluctant nature to multiple biological systems and high catalytic activities and the most advantageous the peroxidase-like activities represented by nanozymes. The surface modification of nanozymes allows efficient drug delivery and increases the efficacy and bioavailability of the drugs. The use of nanozymes has tremendously helped in the field of diagnosis, cancer treatment. Also, perforable progress has been made in generating the alternatives to treat the MDR and XDR infections.

References

[1] V. Salles, S. Bernard, R. Chiriac, P. Miele, Structural and thermal properties of boron nitride nanoparticles, Journal of the European Ceramic Society 32 (2012) 1867-1871. https://doi.org/10.1016/j.jeurceramsoc.2011.09.002

[2] D. Lombardo, M.A. Kiselev, M.T. Caccamo, Smart nanoparticles for drug delivery application: development of versatile nanocarrier platforms in biotechnology and nanomedicine, Journal of Nanomaterials 2019 (2019). https://doi.org/10.1155/2019/3702518

[3] A.K. Barui, R. Kotcherlakota, C.R. Patra, Biomedical applications of zinc oxide nanoparticles, Inorganic frameworks as smart nanomedicines, Elsevier2018, pp. 239-278. https://doi.org/10.1016/B978-0-12-813661-4.00006-7

[4] C.L. Ventola, Progress in nanomedicine: approved and investigational nanodrugs, Pharmacy and Therapeutics 42 (2017) 742.

[5] T.H. Kim, H.H. Jiang, Y.S. Youn, C.W. Park, K.K. Tak, S. Lee, H. Kim, S. Jon, X. Chen, K.C. Lee, Preparation and characterization of water-soluble albumin-bound

curcumin nanoparticles with improved antitumor activity, International journal of pharmaceutics 403 (2011) 285-291. https://doi.org/10.1016/j.ijpharm.2010.10.041

[6] C. Chakraborty, A.R. Sharma, G. Sharma, S.-S. Lee, Zebrafish: A complete animal model to enumerate the nanoparticle toxicity, Journal of nanobiotechnology 14 (2016) 1-13. https://doi.org/10.1186/s12951-016-0217-6

[7] W.N. Souery, C.J. Bishop, Clinically advancing and promising polymer-based therapeutics, Acta biomaterialia 67 (2018) 1-20. https://doi.org/10.1016/j.actbio.2017.11.044

[8] Q. Wu, A. Nouara, Y. Li, M. Zhang, W. Wang, M. Tang, B. Ye, J. Ding, D. Wang, Comparison of toxicities from three metal oxide nanoparticles at environmental relevant concentrations in nematode Caenorhabditis elegans, Chemosphere 90 (2013) 1123-1131. https://doi.org/10.1016/j.chemosphere.2012.09.019

[9] C. Wang, T. Chang, S. Dong, D. Zhang, C. Ma, S. Chen, H. Li, Biopolymer films based on chitosan/potato protein/linseed oil/ZnO NPs to maintain the storage quality of raw meat, Food Chemistry 332 (2020) 127375. https://doi.org/10.1016/j.foodchem.2020.127375

[10] L. Gao, J. Zhuang, L. Nie, J. Zhang, Y. Zhang, N. Gu, T. Wang, J. Feng, D. Yang, S. Perrett, Intrinsic peroxidase-like activity of ferromagnetic nanoparticles, Nature nanotechnology 2 (2007) 577-583. https://doi.org/10.1038/nnano.2007.260

[11] J. Li, C. Shi, X. Wang, C. Liu, X. Ding, P. Ma, X. Wang, H. Jia, Hydrogen sulfide regulates the activity of antioxidant enzymes through persulfidation and improves the resistance of tomato seedling to copper oxide nanoparticles (CuO NPs)-induced oxidative stress, Plant Physiology and Biochemistry 156 (2020) 257-266. https://doi.org/10.1016/j.plaphy.2020.09.020

[12] W. Liu, L. Tian, J. Du, J. Wu, Y. Liu, G. Wu, X. Lu, Triggered peroxidase-like activity of Au decorated carbon dots for colorimetric monitoring of Hg 2+ enrichment in Chlorella vulgaris, Analyst 145 (2020) 5500-5507. https://doi.org/10.1039/D0AN00930J

[13] P. Rai, V.P. Singh, J. Peralta-Videa, D.K. Tripathi, S. Sharma, F.J. Corpas, Hydrogen sulfide (H2S) underpins the beneficial silicon effects against the copper oxide nanoparticles (CuO NPs) phytotoxicity in Oryza sativa seedlings, Journal of Hazardous Materials 415 (2021) 124907. https://doi.org/10.1016/j.jhazmat.2020.124907

[14] Y. Wang, P. Zhang, W. Fu, Y. Zhao, Morphological control of nanoprobe for colorimetric antioxidant detection, Biosensors and Bioelectronics 122 (2018) 183-188. https://doi.org/10.1016/j.bios.2018.09.058

[15] S. Singh, Nanomaterials exhibiting enzyme-like properties (Nanozymes): Current advances and future perspectives, Frontiers in chemistry 7 (2019) 46. https://doi.org/10.3389/fchem.2019.00046

[16] P.K. Tiwari, A.K. Singh, V.P. Singh, S.M. Prasad, N. Ramawat, D.K. Tripathi, D.K. Chauhan, A.K. Rai, Liquid assisted pulsed laser ablation synthesized copper oxide nanoparticles (CuO-NPs) and their differential impact on rice seedlings, Ecotoxicology and environmental safety 176 (2019) 321-329. https://doi.org/10.1016/j.ecoenv.2019.01.120

[17] K. Dai, T. Peng, D. Ke, B. Wei, Photocatalytic hydrogen generation using a nanocomposite of multi-walled carbon nanotubes and TiO2 nanoparticles under visible light irradiation, Nanotechnology 20 (2009) 125603. https://doi.org/10.1088/0957-4484/20/12/125603

[18] D. Böhme, A.G. Beck-Sickinger, Drug delivery and release systems for targeted tumor therapy, Journal of Peptide Science 21 (2015) 186-200. https://doi.org/10.1002/psc.2753

[19] C.Y. Park, J.M. Seo, H. Jo, J. Park, K.M. Ok, T.J. Park, Hexagonal tungsten oxide nanoflowers as enzymatic mimetics and electrocatalysts, Scientific reports 7 (2017) 1-11. https://doi.org/10.1038/s41598-016-0028-x

[20] R. André, F. Natálio, M. Humanes, J. Leppin, K. Heinze, R. Wever, H.C. Schröder, W.E. Müller, W. Tremel, V2O5 nanowires with an intrinsic peroxidase-like activity, Advanced Functional Materials 21 (2011) 501-509. https://doi.org/10.1002/adfm.201001302

[21] S. Liu, F. Lu, R. Xing, J.J. Zhu, Structural effects of Fe3O4 nanocrystals on peroxidase-like activity, Chemistry-A European Journal 17 (2011) 620-625. https://doi.org/10.1002/chem.201001789

[22] A.A. Vernekar, T. Das, S. Ghosh, G. Mugesh, A remarkably efficient MnFe2O4-based oxidase nanozyme, Chemistry-An Asian Journal 11 (2016) 72-76. https://doi.org/10.1002/asia.201500942

[23] Q. Cai, S. Lu, F. Liao, Y. Li, S. Ma, M. Shao, Catalytic degradation of dye molecules and in situ SERS monitoring by peroxidase-like Au/CuS composite, Nanoscale 6 (2014) 8117-8123. https://doi.org/10.1039/c4nr01751j

[24] A. Vinosha, E. Jeronsia, K. Raja, A. christina Fernandez, S. Krishnan, J. Das, Investigation of optical, electrical and magnetic properties of cobalt ferrite nanoparticles by naive co-precipitation technique, Optik 127 (2016) 9917-9925. https://doi.org/10.1016/j.ijleo.2016.07.063

[25] W. Shi, X. Zhang, S. He, Y. Huang, CoFe2O4 magnetic nanoparticles as a peroxidase mimic mediated chemiluminescence for hydrogen peroxide and glucose, Chemical Communications 47 (2011) 10785-10787. https://doi.org/10.1039/c1cc14300j

[26] M. Karimi, A. Ghasemi, P.S. Zangabad, R. Rahighi, S.M.M. Basri, H. Mirshekari, M. Amiri, Z.S. Pishabad, A. Aslani, M. Bozorgomid, Smart micro/nanoparticles in stimulus-responsive drug/gene delivery systems, Chemical Society Reviews 45 (2016) 1457-1501. https://doi.org/10.1039/C5CS00798D

[27] V.P. Torchilin, Multifunctional, stimuli-sensitive nanoparticulate systems for drug delivery, Nature reviews Drug discovery 13 (2014) 813-827. https://doi.org/10.1038/nrd4333

[28] M. Li, G. Zhao, W.-K. Su, Q. Shuai, Enzyme-Responsive Nanoparticles for Anti-tumor Drug Delivery, Frontiers in Chemistry 8 (2020). https://doi.org/10.3389/fchem.2020.00647

[29] Q. Zhou, S. Shao, J. Wang, C. Xu, J. Xiang, Y. Piao, Z. Zhou, Q. Yu, J. Tang, X. Liu, Enzyme-activatable polymer-drug conjugate augments tumour penetration and treatment efficacy, Nature nanotechnology 14 (2019) 799-809. https://doi.org/10.1038/s41565-019-0485-z

[30] A.P. Blum, J.K. Kammeyer, A.M. Rush, C.E. Callmann, M.E. Hahn, N.C. Gianneschi, Stimuli-responsive nanomaterials for biomedical applications, Journal of the american chemical society 137 (2015) 2140-2154. https://doi.org/10.1021/ja510147n

[31] J. Li, L. Feng, L. Fan, Y. Zha, L. Guo, Q. Zhang, J. Chen, Z. Pang, Y. Wang, X. Jiang, Targeting the brain with PEG-PLGA nanoparticles modified with phage-displayed peptides, Biomaterials 32 (2011) 4943-4950. https://doi.org/10.1016/j.biomaterials.2011.03.031

[32] X. Wang, H. Tang, C. Wang, J. Zhang, W. Wu, X. Jiang, Phenylboronic acid-mediated tumor targeting of chitosan nanoparticles, Theranostics 6 (2016) 1378. https://doi.org/10.7150/thno.15156

[33] S. Ruan, M. Yuan, L. Zhang, G. Hu, J. Chen, X. Cun, Q. Zhang, Y. Yang, Q. He, H. Gao, Tumor microenvironment sensitive doxorubicin delivery and release to glioma using angiopep-2 decorated gold nanoparticles, Biomaterials 37 (2015) 425-435. https://doi.org/10.1016/j.biomaterials.2014.10.007

[34] N. Singh, M. Geethika, S.M. Eswarappa, G. Mugesh, Manganese-Based Nanozymes: Multienzyme Redox Activity and Effect on the Nitric Oxide Produced by Endothelial Nitric Oxide Synthase, Chemistry-A European Journal 24 (2018) 8393-8403. https://doi.org/10.1002/chem.201800770

[35] R. Tian, J. Sun, Y. Qi, B. Zhang, S. Guo, M. Zhao, Influence of VO2 nanoparticle morphology on the colorimetric assay of H2O2 and glucose, Nanomaterials 7 (2017) 347. https://doi.org/10.3390/nano7110347

[36] Z. Sun, N. Zhang, Y. Si, S. Li, J. Wen, X. Zhu, H. Wang, High-throughput colorimetric assays for mercury (II) in blood and wastewater based on the mercury-stimulated catalytic activity of small silver nanoparticles in a temperature-switchable gelatin matrix, Chemical Communications 50 (2014) 9196-9199. https://doi.org/10.1039/C4CC03851G

[37] F. Tian, J. Zhou, B. Jiao, Y. He, A nanozyme-based cascade colorimetric aptasensor for amplified detection of ochratoxin A, Nanoscale 11 (2019) 9547-9555. https://doi.org/10.1039/C9NR02872B

[38] G.-W. Wu, S.-B. He, H.-P. Peng, H.-H. Deng, A.-L. Liu, X.-H. Lin, X.-H. Xia, W. Chen, Citrate-capped platinum nanoparticle as a smart probe for ultrasensitive mercury sensing, Analytical chemistry 86 (2014) 10955-10960. https://doi.org/10.1021/ac503544w

[39] W. Zhang, S. Hu, J.-J. Yin, W. He, W. Lu, M. Ma, N. Gu, Y. Zhang, Prussian blue nanoparticles as multienzyme mimetics and reactive oxygen species scavengers, Journal of the American Chemical Society 138 (2016) 5860-5865. https://doi.org/10.1021/jacs.5b12070

[40] A. Zhu, K. Sun, H.R. Petty, Titanium doping reduces superoxide dismutase activity, but not oxidase activity, of catalytic CeO2 nanoparticles, Inorganic chemistry communications 15 (2012) 235-237. https://doi.org/10.1016/j.inoche.2011.10.034

[41] H. Wang, C. Liu, Z. Liu, J. Ren, X. Qu, Specific oxygenated groups enriched graphene quantum dots as highly efficient enzyme mimics, Small 14 (2018) 1703710. https://doi.org/10.1002/smll.201703710

[42] C. Xu, W. Bing, F. Wang, J. Ren, X. Qu, Versatile dual photoresponsive system for precise control of chemical reactions, ACS nano 11 (2017) 7770-7780. https://doi.org/10.1021/acsnano.7b01450

[43] Y. Gao, K. Chen, J.-l. Ma, F. Gao, Cerium oxide nanoparticles in cancer, OncoTargets and therapy 7 (2014) 835. https://doi.org/10.2147/OTT.S62057

[44] S. Santra, S.D. Jativa, C. Kaittanis, G. Normand, J. Grimm, J.M. Perez, Gadolinium-encapsulating iron oxide nanoprobe as activatable NMR/MRI contrast agent, ACS nano 6 (2012) 7281-7294. https://doi.org/10.1021/nn302393e

[45] J. Mu, L. Zhang, M. Zhao, Y. Wang, Catalase mimic property of Co3O4 nanomaterials with different morphology and its application as a calcium sensor, ACS applied materials & interfaces 6 (2014) 7090-7098. https://doi.org/10.1021/am406033q

[46] K. Sobańska, P. Pietrzyk, Z. Sojka, Generation of reactive oxygen species via electroprotic interaction of H2O2 with ZrO2 gel: ionic sponge effect and pH-switchable peroxidase-and catalase-like activity, ACS Catalysis 7 (2017) 2935-2947. https://doi.org/10.1021/acscatal.7b00189

[47] V. Baldim, F. Bedioui, N. Mignet, I. Margaill, J.-F. Berret, The enzyme-like catalytic activity of cerium oxide nanoparticles and its dependency on Ce 3+ surface area concentration, Nanoscale 10 (2018) 6971-6980. https://doi.org/10.1039/C8NR00325D

[48] W. Zhang, J. Dong, Y. Wu, P. Cao, L. Song, M. Ma, N. Gu, Y. Zhang, Shape-dependent enzyme-like activity of Co3O4 nanoparticles and their conjugation with his-tagged EGFR single-domain antibody, Colloids and Surfaces B: Biointerfaces 154 (2017) 55-62. https://doi.org/10.1016/j.colsurfb.2017.02.034

[49] X.-Q. Zhang, S.-W. Gong, Y. Zhang, T. Yang, C.-Y. Wang, N. Gu, Prussian blue modified iron oxide magnetic nanoparticles and their high peroxidase-like activity, Journal of Materials Chemistry 20 (2010) 5110-5116. https://doi.org/10.1039/c0jm00174k

[50] J.L. de Beer, I. Bergval, A. Schuitema, R.M. Anthony, M. Fauville-Dufaux, B.E. Ferro, V. Ritacco, J. van Ingen, A. Zomer, D. van Soolingen, A unique mutation in the rpoC-gene exclusively detected in Mycobacterium tuberculosis isolates of the largest cluster of multidrug resistant cases of the Beijing genotype in Europe, Molecular typing of Mycobacterium tuberculosis complex 105.

[51] A. MacNeil, P. Glaziou, C. Sismanidis, A. Date, S. Maloney, K. Floyd, Global epidemiology of tuberculosis and progress toward meeting global targets-worldwide, 2018, Morbidity and Mortality Weekly Report 69 (2020) 281. https://doi.org/10.15585/mmwr.mm6911a2

[52] D.L. Clemens, B.-Y. Lee, M. Xue, C.R. Thomas, H. Meng, D. Ferris, A.E. Nel, J.I. Zink, M.A. Horwitz, Targeted intracellular delivery of antituberculosis drugs to Mycobacterium tuberculosis-infected macrophages via functionalized mesoporous silica nanoparticles, Antimicrobial agents and chemotherapy 56 (2012) 2535-2545. https://doi.org/10.1128/AAC.06049-11

[53] H.H. Gustafson, D. Holt-Casper, D.W. Grainger, H. Ghandehari, Nanoparticle uptake: the phagocyte problem, Nano today 10 (2015) 487-510. https://doi.org/10.1016/j.nantod.2015.06.006

[54] H. Sung, J. Ferlay, R.L. Siegel, M. Laversanne, I. Soerjomataram, A. Jemal, F. Bray, Global cancer statistics 2020: GLOBOCAN estimates of incidence and mortality worldwide for 36 cancers in 185 countries, CA: a cancer journal for clinicians 71 (2021) 209-249. https://doi.org/10.3322/caac.21660

[55] K. Bukowski, M. Kciuk, R. Kontek, Mechanisms of multidrug resistance in cancer chemotherapy, International journal of molecular sciences 21 (2020) 3233. https://doi.org/10.3390/ijms21093233

[56] R. Vankayala, K.C. Hwang, Near-infrared-light-activatable nanomaterial-mediated phototheranostic nanomedicines: an emerging paradigm for cancer treatment, Advanced Materials 30 (2018) 1706320. https://doi.org/10.1002/adma.201706320

[57] S. Thangudu, C.-H. Su, Peroxidase Mimetic Nanozymes in Cancer Phototherapy: Progress and Perspectives, Biomolecules 11 (2021) 1015. https://doi.org/10.3390/biom11071015

[58] A. Manke, L. Wang, Y. Rojanasakul, Mechanisms of nanoparticle-induced oxidative stress and toxicity, BioMed research international 2013 (2013). https://doi.org/10.1155/2013/942916

[59] M.L. Circu, T.Y. Aw, Reactive oxygen species, cellular redox systems, and apoptosis, Free Radical Biology and Medicine 48 (2010) 749-762. https://doi.org/10.1016/j.freeradbiomed.2009.12.022

[60] L. Zeng, Y. Han, Z. Chen, K. Jiang, D. Golberg, Q. Weng, Biodegradable and Peroxidase-Mimetic Boron Oxynitride Nanozyme for Breast Cancer Therapy, Advanced Science (2021) 2101184. https://doi.org/10.1002/advs.202101184

Nanomaterial-Supported Enzymes   Materials Research Forum LLC
Materials Research Foundations **126** (2022) 215-239   https://doi.org/10.21741/9781644901977-8

Chapter 8

Biomedical uses of Enzymes Immobilized by Nanoparticles

Syed Raza Ali Naqvi[1]*, Noman Razzaq[1], Tahseen Abbas[1], Ameer Fawad Zahoor[1], Matloob Ahmad[1], Amjad Hussain[3], and Sadaf Ul Hassan[4]

[1]Department of Chemistry, Government college university Faisalabad.

[2]Department of Chemistry, COMSATS University Islamabad, Abbottabad Campus, Abbottabad, Pakistan

[3]Department of Chemistry, University of Okara, Okara-Pakistan

[4]Department of Chemistry, School of Sciences, University of Management and Technology, Lahore Campus, Pakistan

* drarnaqvi@gmail.com

Abstract

Immobilized enzymes are now a significant and appropriate area of modern technologies. Immobilization of enzymes on nanoparticles (NPs) especially magnetic nanoparticles (MNPs) not only increase the stability of the enzymes by protecting the active site but also facilitates the separation mode. Immobilized technology is considered effective in context of running cost to exercise immobilized enzymes technique. Nowadays, variety of magnetic nanoparticles are available such as chitin-chitosan magnetic nanoparticles, Fe_3O_4 magnetic nanoparticles, bacteriophages T^4 capsid novozym-435 etc. which are quite fit for loading enzymes and to use fruitfully. The main focus in this piece of work is that how immobilized enzymes are helpful in different biomedical uses and what kind of enzymes and nanoparticles could be hyphenated to take advantage in health care sectors. Different method of enzymes immobilization will also be discussed in details including both physical methods and chemical methods of loading enzymes on nanoparticles.

Keywords

Enzymes, Magnetic Nanoparticles, Immobilized Enzymes, Applications

Contents

1.   Introduction

Enzymes are the major accelerator for almost all biochemical reactions occurring inside or outside of the cells. Those biocatalysts are proteins or a glycoproteins macromolecule with varied number of active sites to normalize its function. They are synthesized in living cells. Activities of enzymes widely impart a great role in the metabolic reactions in the living system particularity in animal and human beings [1-3]. Enzymes are the biocatalysts bearing high sensitivity by nature; therefore, they work at a limited range of temperature and ph. Beyond the limit, temperature and pH badly affect the structure and activity of enzymes due to which they fail to do their job in physiological and industrial processes. Nowadays, technology is progressing day by day and the role of enzymes is increasing exponentially. Plant and microbial enzyme have found its enormous application in the world of agriculture, pharmaceuticals, cosmetics, textiles, paper, environment, biofuel and food industry. However, as recovery is associated with inorganic catalysts in chemical reactions; in contrary, enzymes cannot be recovered. For example, in maximum enzymatic industrial processes, soluble enzymes are used. These enzymes got deactivation in drastic conditions due to its limited range of sensitivity which make the situation impossible to recover, due to which the cost of the product increased significantly. These disadvantages of enzymes, triggered the development of new system for maintaining the enzyme active, intact and recoverable in wide range of reaction conditions. In order to take maximum advantage of existed stuff of enzyme by keeping the enzymes active, intact and recoverable under multiple reaction conditions, enzyme immobilization technology was adopted [4, 5]. Immobilization method itself was first developed in 1916 and it expand to different sectors especially biocatalyst field. For enzyme immobilization different supporting materials were practiced including organic & synthetic polymers, silica gels, hydrogels, mesoporous materials and NPs. Among these supports, NPs enabled the enzymes to work in somewhat harsh conditions. The technique was remained very successful and immobilized enzymes by nanoparticles showed a good profile of advantages [6]. Stability and formation of nanoparticles for enzyme immobilization plays an important role in order to get fruitful results. It is because, properties of nanoparticles are definitely depending on the development as well as their construction methods [7]. Symmetry, particle size, crystallization and particle distribution ratio are the phenomenon's that can be controlled easily as compare to other supports [8].

In recent years, nanoparticles are not only used for immobilization but also used in catalysis, biomedicines, micro and nano-fluids, for ecological remedies, drug-delivery, optical fibers and as data storage [9, 10]. However, in present era the acceleration of using the Nanoparticles as a carrier material for the immobilization of proteins, drugs, enzymes and antibodies is gaining the momentum [11]. For example, the activity of lipase for the first time increased many times by immobilization method using MNPs of Fe_3O_4 as supporting material, which indicates that in immobilization technique the nature of the carrier material and method to load the enzyme play critical role to improve the enzyme activity and stability [12, 13].

These particles have a major impact in biotechnology and multiple areas of biomedicines. These immobilized particles can be directly use to heal a disease or to modify cell and molecule on to the targeted areas. Immobilization by NPs is a significant area in biotechnology which offer several different kinds of supports such as nanofibers, nano-flowers and magnetic nanoparticles to hold enzymes for particular function in living system [14-16]. Presently, various NPs are being used for immobilization of enzymes [17]. However, MNPs appear more effective because of its broad-spectrum.

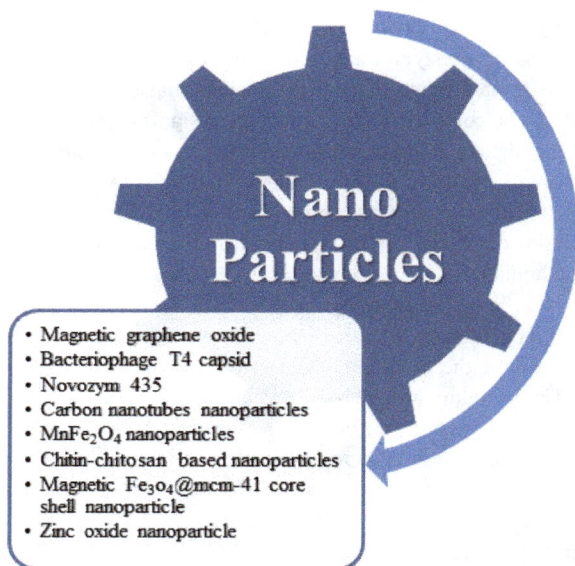

Nano Particles

- Magnetic graphene oxide
- Bacteriophage T4 capsid
- Novozym 435
- Carbon nanotubes nanoparticles
- $MnFe_2O_4$ nanoparticles
- Chitin-chitosan based nanoparticles
- Magnetic Fe_3O_4@mcm-41 core shell nanoparticle
- Zinc oxide nanoparticle

Figure 1: Nanoparticles used for enzymes immobilization

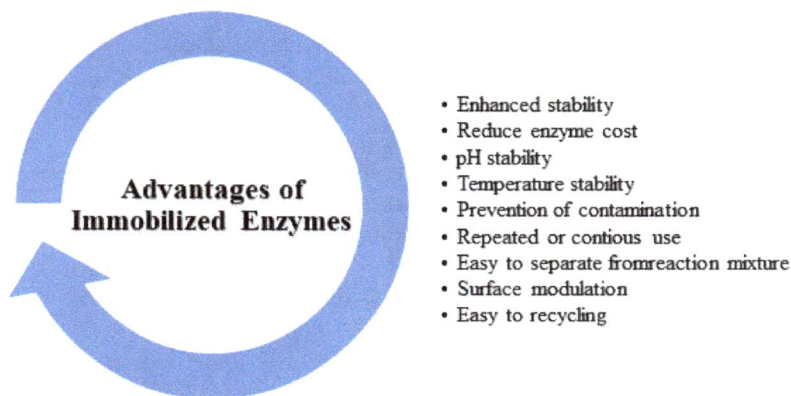

Advantages of Immobilized Enzymes

- Enhanced stability
- Reduce enzyme cost
- pH stability
- Temperature stability
- Prevention of contamination
- Repeated or contious use
- Easy to separate fromreaction mixture
- Surface modulation
- Easy to recycling

Figure2: Advantages of immobilized enzymes

These properties include enormous surface to volume ratio which is too much important for immobilization point of view, large surface area, loading capacity and specifically magnetic/superparamagnetic property. Magnetic orsuperparamagnetic nanoparticles used as a useful carrier because MNPs contains magnetic properties, ease to separate under the external magnetic field [18]. Most commonly used MNPs are Fe_3O_4 and Maghemite due to very low toxicity levels. Hybrid technology made a lot of the materials which are affordable for immobilization technology. So organic and various inorganic substances are also used for enzymes immobilization. For immobilization physiochemical properties, pore size, surface chemistry should be observed. MNPs carriers comprises three major parts. These parts are called functional parts which consists of a magnetic core, surface area and outer coating [19].Transition metal complexes sometimes are used for binding the numerous catalytic substances. Binding of catalytic particles to MNPs for its multiple applications offer pure production of NPs loaded with enzymes, therefore, no additional purification step is required [20]. Recent development in using MNPs as supporting agent, core-shell structured magnetic nanoparticles are being used. Figure 1 shows the reported nanoparticles used for enzyme immobilization. The major advantage associate with its use are their low toxicity and chemically adaptable surface. The process, immobilized enzymes in conclusion, offer Stability at extended range of temperature and pH and great efficiency as compare to mobilized enzymes. What actually the major advantages of enzyme immobilized process can be seen in the following Figure 2 [8].

2.   Enzymes immobilization methods

Various methods have been adopted for immobilization of enzymes. Enzyme immobilization techniques or method includes *adsorption binding, covalent binding, entrapment, cross linked with/without support and affinity immobilization* [21]. Enzyme immobilization by force of affinity between enzyme and carrier nanoparticle, a physical adsorption works well in some particular applications. The adsorption may take place either through hydrogen bonding, or weak van der Waals interactions [22]. Ionic bonding is strongest as compared to physical binding because it avoids the enzymes leaching from the surface. Formation of covalent bond take place by the reaction between the functional group of unchanged enzymes. It comprises the example of enzymes such as lysine, aspartic acid, glutamic acid and cysteine. The enzymes immobilization methods are shown graphically in Figure 3. Chemical adsorption sometime needs prior modification of nanoparticle substrate surface to make it enable for chemical bonding with enzymes. After modification enzymes are ready to attach at the surface of carrier nanoparticle. Modification can be proceeded either by modifying the functional group of nanoparticle or by attaching the functional group of nanoparticles with some coupling agents [23]. Different functional groups and sometimes polymers are also coated over enzyme surface which offer functional stability of enzyme activity [24].

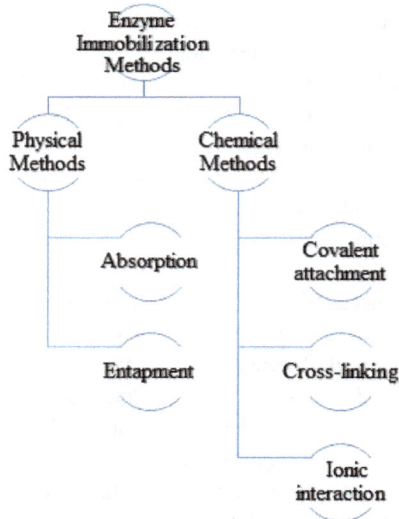

Figure 3: Classification of enzyme immobilization methods

3.   Choice of supports

For the preparation of immobilized enzymes using nanoparticle, both carries material and properties of enzymes plays a significant role. Demonstrative of carrier's material should be matched with the properties of enzymes because performance is totally based on the characteristic of both carrier materials and enzymes. Biochemical, mechanical, chemical and kinetic properties of immobilized enzymes depend on the interaction between the support material and enzymes so order to get required results. We classified the support in to organic, inorganic or biopolymer which totally depends on the chemical composition. Organic support can be further divided into synthetic and natural polymer. Properties of supporting material consist of resistance to compression, hydrophilicity, inertness in the direction of enzymes ease of derivitization, biocompatibility, fight to microbial attack, and availability at low cost, are involved [8].

3.1   Entrapment

It is based on the constriction of the enzymes into the polymeric network of organic or inorganic such as polyacrylamide and silica-based material, respectively. Hollow fiber or microcapsule which basically called membrane devices can also be used polymeric network following the rule of diffusion to allow the reactant and product to diffuse freely while dodging the passage of biocatalyst. Immobilization of enzymes done by this method are effective and low cost. Due to small pore size of matrix, it decreases mass transfer of substrate to enzyme. Sometimes enzyme is deactivated during immobilization. This problems is solved by the addition of cross-linking agent [25]. The entrapment immobilization method is mostly used. Entrapment method permits the substrate and product while retaining the enzyme. Different methods such as gel trapping, fiber entrapping or microencapsulation involved in this method. This method is carried out while using immobilized whole cells. It usually needs the development of polymeric medium in the presence of enzymes.it is a low-cost method. Different approaches must be adopted in order to avoid the arrogation of enzymes. Different media that are being used for entrapment methods are chitosan, silica, chitin-chitosan beads, rubber, collagen poly acrylamide, polyurethane, agar, polyvinyl alcohol, gelatin and cellulose triacetate. Gel/fiber entrapping, micro capsulation are the methods which are used for immobilization of enzymes. these useful methods allow substrate and products to pass easily while enzymes are retained in the networks means polymeric networks. It may be helpful to avoid the leaching of enzymes and gives extra stability and enzymatic reactions are carried out in a well performed manner. High catalytic activity should be maintained and in order to prevent the movement of macromolecules diffusion barrier may be helpful. Instead of using polymeric networks, researchers recently developed the MOFs for immobilized enzymes.in this new method, to entrap the enzyme we should prefer to use Zeolitic imidazolate networks (ZIF-8). We added polyvinylpyrrolidone (PVP) which act as a stabilizer it is than protected and then mixed with zinc nitrate to synthesis MOFs immobilized enzyme [6].

3.2   Crosslinking

This method involves the bifunctional group which act as a cross linking agent. The most commonly cross-linking agent that are used for enzymes immobilization is glutaraldehyde. Enzyme's immobilization that has done by using crosslinking consisting of major physical forces such hydrophobic interactions, ionic interaction and most important van der walls interactions. But under high condition of temperature these interactions become too weak to be detached. Addition of precipitants such as acetone, ethanol along with glutaraldehyde forming diamine bonds. Some additives are also added in order to increase the stability of enzymes. Moreover, crosslinking aggregates are also used in order to enhance the functionalities. It is totally based on intermolecular interactions. It is the type of chemical method. This method comprises the cross-linked of enzymes to the support materials using bifunctional components. For high performance applications, covalent bonded attachment to nanoparticles are used. But enzyme catalytic activity diminishes by this method to some extent. However, coupling agent for immobilization process preserve the enzyme activities promisingly and therefore, this immobilization method used mostly. Among crosslinking method, cross-linked enzymes crystals (CLES) provide a stable configuration, impact a high catalyst-to-weight ratio and increase the catalytic activity. But CLES limited to only certain enzymes. so we introduces cross-linked enzymes aggregate (CLEA) to overcome this drawback because it can applied to wide range of proteins [26].Glutaraldehyde is used mostly for the preparation of cross-linked enzyme aggregate (CLEAs) as shown in Figure 4, enzymes are crosslinked to another enzymes. Firstly, the precipitants for aggregation process and afterwards using bifunctional reagents for the preparation of CLEAs. This is done in order to protect the active site of enzyme. The researcher used 75% ammonia sulphate as a precipitant and glutaraldehyde as across linker.

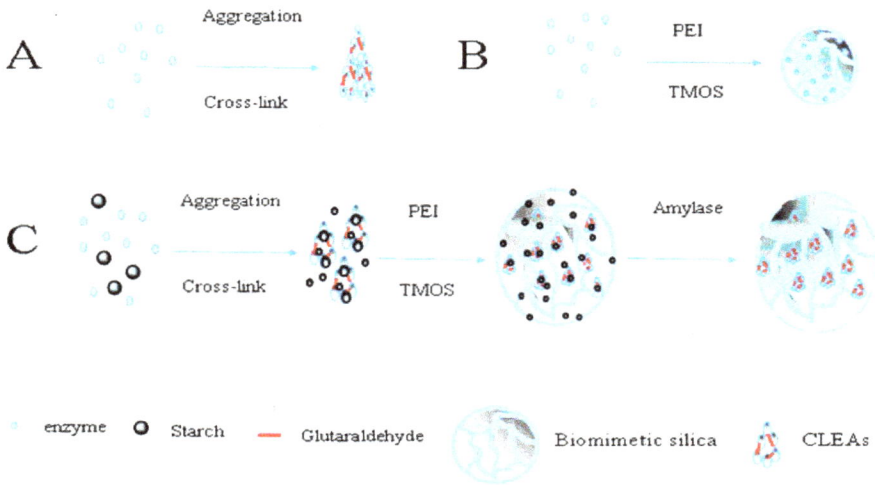

Figure 4: Preparation of CLEAs

3.3   Covalent attachment

It is another important chemical method for immobilization of enzymes. For immobilization of enzymes the most conventional method is covalent attachment. Formation of covalent bond is due to the reaction between the support material that is used as carrier and enzymes. amino acids such as lysine, aspartic acid, glutamic acid and lysine which mostly follow covalent attachment. Many functional groups of amino acid also follow the application of covalent attachment such as phenolic group, hydroxyl group, indolyl, imidazole, carboxylic group and amino group etc.in order to prevent leakage of enzymes covalent attachment is preferred thus this binding improving the stability of immobilized enzymes [27]. This attachment is also helpful in order to activate the sites of enzymes. Many methods such as bromide cyanogen, diazotization, glutaraldehyde coupling have been developed by using the mode of covalent attachment. Among these methods glutaraldehyde coupling method is most common and in majority cases it was used [6].Multipoint covalent attachment (MCA)of various enzymes to carrier materials increasing stabilizing effect for example stabilizing factor increases up to 10000-fold by immobilized on glyoxal agarose through MCA. Multipoint covalent attachment provides rigidification to enzymes that enhance structure stability. MCA is somewhat difficult to achieve so we mostly prefer simple covalent attachment [28].

3.4   Adsorption

It is one of the simplest methods for enzyme immobilization method. Physical adsorption, ion adsorption and affinity adsorption are the typical types of this kind of immobilization. We used resins mostly solid support resin such as cation and anion exchange resin. Enzymes are adsorbed on the surface of supports material by interaction of these resins for enzymes. Sometimes hydrogen bonding also involves but Vander Waals's interaction is most operative. For example, lipase was immobilized on the resins of phenyl-Sepharose, laccases on the surface and into the pores of Zr-metal. Adsorption method is very helpful in cost perspective point of view [29]. In this method no additional substance, additives and coupling agent required for stability. Moreover, the activity of enzymes gives fruitful results. The only demerit while using adsorption method is a little interaction between enzyme and carriers' materials, due to which the probability of enzyme leaching exist [30]. The summary of functional groups opperated in immobilization of enzymes by physical and chemical methods are shown in Table 1.

Table 1: Involvement of functional groups in enzymes immobilization processes

Method of immobilization	Functional group	Binding	Reactive group on enzyme
Van der Waals	Alkyl	Hydrophobic	Hydrophobic areas on the surface of lipase
Ionic interaction	Trialkyl amine carboxylate	Ionic adsorption	Negatively charged a. a
Covalent bonds	Epoxy Amino Diol	Covalent bond	Primary amines Nucleophiles groups Primary amines(terminal)

4.   Carrier bound method: general concept

Immobilization of enzymes done by classical method involve the formation of chemical interaction between the solid support and enzyme. Supports can be chosen on the basis of adsorption or covalent bonding immobilization, that have already been discussed in details. Immobilization generally occur in the aqueous medium which contains the enzymes in the dissolved form while the carrier material which is in solid always in suspended form. In case of the aqueous buffer ionic strength should be adjusted. It encourages the barrier of protein on to the solid transporter. Majority of the enzymes that are used for immobilization have the hydrophilic surface [31]. Therefore, the removal of water molecule from the protein surface and formation of new polar interaction can occur relatively of low ionic strength. Selection of above strategy, the following method requires a number of factors such as activity of enzymes, enzyme deactivation, cost, reuse, renewal characteristic and properties of enzymes. We prefer affinity immobilization because of its specific interactions. It gives good control on immobilization and restricted to minimum structure changes. Moreover, purification of enzymes must be done by using the method of affinity immobilization. Due to this affinity immobilization is advantageous over the other methods. choice of the supporting materials is another essential procedure. In recent years, nanoparticles illustrated growing interest because their fundamentally large surface area for enzyme effectiveness, even as immobilizing enzymes onto planar surfaces commonly decreased the activity however, their application is hindered by means of high cost [32]. Researchers are now developing simple and cost affective enzymes for immobilization. High specificity and catalytic effectiveness of enzymes in biological reactions makes it very significant [33]. However, a lot of limitations includes high cost of preparations, refinement, separation, stability, severe circumstances of environment comprise high temperature, ph. In order to overcome the above drawbacks development of artificial enzymes now enhanced. Now nanomaterials with enzymes are called nanozymes entered in the next generation of enzyme impersonators. These nanozymes get the importance with the unexpected discoveries of magnetic nanoparticles. The smooth surface of nanoparticles provides more space for immobilization which enhanced both the enzymatic and catalytic activities. Moreover, physicochemical possessions of nanoparticles along with the substantial improvement made in the few years by nanotechnology, biotechnology, catalytic technology to increase new enzymatic activities. Mostly nanoparticles belong to the nano-alloys that are coated with special kind of molecules. Sometimes we do modifications in order to increase the function of enzymes.it may also be helpful to improve physiochemical properties such as corrosion, oxidation, and toxicity of nanoparticles [34]. For example, magnetic nanoparticle with avidin-biotin which act as a linker to the MNPs increase the stability of immobilized enzymes to several months. Magnetic nanoparticles also layered with molecules mostly conjugated molecules like micelles, liposomes, polymeric coating and core-shell- structures. These all operations done just only to increase the stability of enzymes [35]. In the following section we will discuss the application of immobilized enzymes in details.

5. Degradation of dye pollutants

Some dyes such as acid blue 45, orange G shows many applications in the fields of food, textile, paper, etc. But these dyes contain pollutants that are injurious to human health and the organisms that are living in aquatic environment. The industrialist is applying chemical, physical and biological treatment method but these are the classical methods which have a lot of shortcomings. However immobilized enzymes method for the treatment of pollutants especially in the treatment of dyes, offer many advantages over classical methods. Chloroperoxidase (CPO) is the most common enzyme used for degradation [36]. But it has lack of stability and recovering but still it has widespread applications in wastewater bioremediation. To resolve the issue of constancy and recycling, the immobilization of CPO and Glucose oxidase (GOx) enzymes were practiced by loading over magnetic nanoparticle surface. The enzymes were loaded over the surface of Magnetic graphene oxide nanoparticles. Its catalytic activity increases up to almost 90%. It shows maximum activity in order to degrade the pollutants present at 40°C. Almost by using the method of co-immobilization of CPO along with GOx by using graphene oxide as nanoparticles almost decolorization of6 dyes achieved. It is also widely used in environment applications. An alternative method is also applied for decolorization of dyes and also used for the purification of textile waste water. The enzyme that used for this process is Peroxidase enzyme that is isolated during the purification of industrial waste water treatment. Magnetic nanoparticle that are used for immobilization is Fe$_3$O$_4$. So, in order to increase the stability under pH and temperature changing, Fe$_3$O$_4$ used glutaraldehyde as a functionalized member during immobilized with peroxidase. Due to this catalytic activity increasing up to 100 folds and reusing it again and again. It is widely used for the decolorization as well as degradation of waste water which is abundant in green, red or reactive red azo dye pollutants. Sometimes we used CuFe$_2$O$_4$ as a magnetic nanoparticle instead of Fe$_3$O$_4$ it is used for immobilization of laccase enzymes to degrade an azo dye [37]. Researchers revealed that immobilized biocatalyst plays a maximum role for decolorization of different dyes under the normal conditions. Their estimated efficiency is about 94%.it can be stated that magnetic nanoparticles along with immobilized enzymes also play a role for decolorization of dyes [35].

6. Fe$_3$O$_4$ along with L-asparaginase

L-asparaginase is the 1st approved bacterial enzymes that is used for treatment of cancer. Aspartic acid and ammonia are formed after hydrolyzed these enzymes. L-asparagine synthetase (ASNS) that is present in normal cell catalyze the reaction between L-aspartate and L glutamine for the synthesis of L-asparaginase (ASN). Now this enzyme is widely used to treat acute lymphoblastic leukemia (ALL) in children across the world. The survival rate of patients 54% during the year(1975-1977) and now 90% [38]. In recent reports, Fe_3O_4 nanoparticle were used for enzyme immobilization. The preference of Fe_3O_4 nanoparticles is due to immobilization developmental advantages such as non-toxicity, enormous surface area, mechanical strength, mechanical properties and the surface area of Fe_3O_4 can easily adapted with a lot of functional groups which highly compatible to application requirements. Chemicals needed to synthesize Fe_3O_4 MNPs includes sodium hydroxide (NaOH), ferric chloride ($FeCl_3.6H_2O$) and ferrous chloride tetrahydrate ($FeCl_2.4H_2O$) [39].

However, surface modification is a successful strategy to improve or to make an enzyme active in some particular applications; such as 3-choloropropyltrimethoxysilane (CPTMS) is being used for surface modification in order to immobilized L-asparaginase on the carrier material. Surface modification is also carried out by coating the nanoparticle surface such as Fe_3O_4 coated by MCM-41 which is the mesoporous structure materials. Due to low toxicity, particle size, pore size, morphology and most important thermal stability it also serves as a support material for drug delivery system. Sometime this mesoporous material act as a catalyst and sometimes it acts as template for the synthesis and purification of other versatile materials. In some modern reports Fe_3O_4@MCM-41 core shell nanoparticles now used in advance levels for immobilization of different enzymes. But it may helpful for modification of very small molecules with functional groups such as epoxy, thiol and amino using chemical (covalent) binding. It is because it attached the carrier material along with enzymes so firmly also preventing leaching of enzymes. CPTMS is an organic agent and only react with the giant molecules such as proteins, carbohydrates, lipids, fatty acids and enzymes. It also acts as a carrier for immobilization of enzymes [40]. It is used because it increases the yield and reusability of enzymes. In order to treat acute lymphoblastic leukemia which is most common in children we used L-Asparaginase. It is an anti-cancer medicine mostly used in chemotherapy. Hydrolyzation of L-Asparaginase into L-Asparagine and then catalysis into L-aspartate and ammonia. Synthesis of proteins for normal as well as for cancer cell we used amino acid L-asparagine. It is produced by L-asparaginase synthetase enzyme. But in cancerous cell there is lack of this enzyme. So, hydrolyzed L-asparagine in order to prevent the division of cancerous cells is preferred. This must be done by immobilization of L-asparaginase by using Fe_3O_4@MCM-41 as a carrier material. FTIR spectroscopy play a role in order to check the chemical composition. The spectra for Fe_3O_4@MCM-41 gives the vibrational value at 1072 and 1620 cm^{-1} it shows 2 new peaks after modification with CPTMs. Thermogravimetry analysis must be done in order to check the stability up to which temperature it survives before immobilization. The efficiency must be almost 70% that is detectable. Optimum ph ranges from 4 to 10 and optimum temperature value from 25 to 70°C [41].

For the improvement of stability and sensitivity L-asparaginase now immobilized on various support materials such as fructose polymer levan, hydrogel magnetic nanoparticle, nanofibers of polyaniline, glutaraldehyde activated silica gel, magnetic nanoparticles functionalized with poly(2-vinyl-4,4-dimethylazlactone) and, magnetic poly(HEMA-GMA) nanoparticle. L- asparaginase has some disadvantages besides its threpautic effect. It causes allergic reactions, having a very short half life period. These disadvantages now overcomed [42].

7.   Chitin and chitosan support material for immobilization

Chitin is a naturally abundant polymer composed of glucosamine and N-acetylated glucosamine. Chitin existed as α and β allomorph but α allomorph is most abundant in nature and found in cell wall of yeast and fungi. It is a semicrystalline biopolymer consisting of microfibrils that are tightly bonded to one another in an extensive network by covalent bond. It is used as a support material for enzymes immobilization due to its biodegradability. Chitosan is a derivative of chitin and found in the skeleton of crustaceans. Chitin is converted into chitosan through enzymatic deacetylation or by chemical treatment. Chitosan in the form of beads are more effective for enzymes immobilization [30]. Performance of enzymes specially immobilized enzyme depends on the properties of the material's surface at which enzyme is loaded. We use chitin and chitosan-based support materials because of its stability and cost-effective bioprocessing. Chitosan contains, having a lot of functional groups, its non-toxicity, its bridgeable nature makes it good supporting material. Moreover, chitosan is the second richest biopolymer in nature. The chief source of chitin is mainly exoskeleton of marine crustaceans specially crabs. Derivative of chitin is chitosan which is formed by the process of deacetylation. Chitosan is made of randomly deacetylated unit(d-glucosamine) and acetylated unit(N-acetyl-d-glucosamine). It consists of a lot of the exclusive properties such as physicochemical properties, biodegradability, non-toxicity and most important attachment properties. Entrapment must be done through crosslinking especially ionic cross linking. Most of the biological functionalities such as bone formation, antitumor, central nervous system depressant and antimicrobial can be improved by using chitosan. Thus, chitin is a useful low-cost biopolymer. So that's why we use chiton and chitosan base support material for enzyme immobilization. We use immobilized enzymes by using chitin as a carrier material because it withstands the harsh conditions due to verity of functional groups such as amino, hydroxyl and polyamine. Such functional groups of chitosan impart extra stability to enzymes as well as carrier materials [43]. It also enables the binding between the carrier material and enzymes so functional groups play a very important role in any standpoint of view for immobilization. For example, amino group of chitosan may be a very helpful in

Nanomaterial-Supported Enzymes   Materials Research Forum LLC
Materials Research Foundations **126** (2022) 215-239   https://doi.org/10.21741/9781644901977-8

order to bind the glutaraldehyde via covalent bond with a protein [44]. Many enzymes include isomerase, lactase, glucose oxidase, acid phosphatase and chymotrypsin are immobilized by using chitin. Chitin and chitosan support-based materials have wider applications in different fields which are given below.

7.1 Biomedical applications

Table 2 is a brief summary of enzyme immobilization application in clinical setups; which indicate the important role of chitosan as a carrier material for several enzymes. Table 3 indicate the use in food and agriculture sectors. It provides extractability to enzymes so it works at optimum temperature and ph. Serrati peptidase enzyme immobilized on chitosan medium shows highest protein loading. So, both the enzyme and carrier materials have their importance. Intrinsic protein also immobilized with chitosan in in order to study the activity of different tissues [45]. It reduced the operational time of blood clot by almost 60%. Researcher preferred protein immobilization chitosan coated magnetic nanoparticle and the efficiency degree inclined up to several folds when we used nanoparticle other than chitosan support based [46]. Researchers are looking for the development of new method for the delivery of insulin. In this regard researcher preferred glutamic acid that is attached with chitosan modified trimethyl functional group. It necessary for strong attachment.so the overall summary is that chitin and chitosan support-based material is an excellent choice for the immobilization enzyme and drug delivery system especially oral drug delivery. That's why we called it a universal biopolymer [47].

Table 2: Biomedical application of immobilized enzymes using chitin-chitosan support material.

Sr. No	Enzymes	Immobilization method	Applications
1	Insulin	Trypsin and goblet cell attached chitosan nanoparticle	Insulin preparation for oral insulin delivery
2	Urease	Chitosan beads	Urea analysis in sample
3	Trypsin	Magnetic nanoparticles functionalize with chitosan	For protein degradation
4	Oxalate oxidase	Mucin/chitosan	Biosensor for oxalate determination
5	Glucose oxidase	Chitosan-ferrocene	Blood glucose biosensor
6	Serrati peptidase	Chitosan	Drug targeting and anti-inflammatory agent
7	Lactate dehydrogenase	Carbon nanotubes-nanoparticles	Lactate biosensor

Table 3: Application of immobilized enzymes in food and agriculture industry.

Sr. No	Enzyme	Support material	Applications
1	Pepsin	Chitosan beads	Saccharification of dextrin to glucose, whipping qualities to proteins, estimation of protein in nondairy
2	Amylase	Chitosan beads	Wide application in food, fermentation and pharmaceuticals
3	Lipase	Chitosan flakes	Flavor industry and detergents
4	Proteases	Chitosan in the form of powder	Textile industry and leather industry
5	Papain	Chitosan beads	Feeds, meat tenderization and textile industry
6	Acid phosphatase	Chitosan beads	Used in labelling of proteins, used in dephosphorylation of nucleic acid and for production of enzyme-based biosensor
7	Amino acylase	Chitosan based alginate beads	Used in vast production of phenylalanine
8	Invertase	Chitosan solution	For the production of bakery products
9	Laccase	Chitosan nanoparticles and magnetic chitosan microsphere	For imparting color enhancement in tea, for cork modification, brewing, fruit juice processing and beer stabilization

8.   Zinc oxide nano-particles

Zinc oxide nanoparticles are unique particles. They are semiconductor in nature. These nanoparticles are prepared from different methods such as hydrothermal methods, electrochemical deposition methods, microwave combustions, thermal decomposition, sol gel method, combustion method and electrophoretic depositions [10]. Global production of ZnO nanoparticles is 0.1-1.2 million ton per year. It can widely use in cosmetics, beauty care products but use of ZnO nanoparticles to human being is still limited. Immobilized enzymes using ZnO nanoparticles enter to the target via inhalation, skin contact and by digestion. Two most important parameters influence on ZnO NPs. One is the chemistry of media including ph and other is physiological properties of zinc oxide ion after released from ZnO [48].They are widely used in biosensors, pigments and food additives. Zinc oxide nanoparticles are also known permanent white, oxo zinc, ketozinc and ox datum.

Several methods are developed for the preparation of ZnO NPs but famous one are precipitation method, wet chemical synthesis method, solid-state pyrolytic method, biosynthesis and sol-gel method [49].

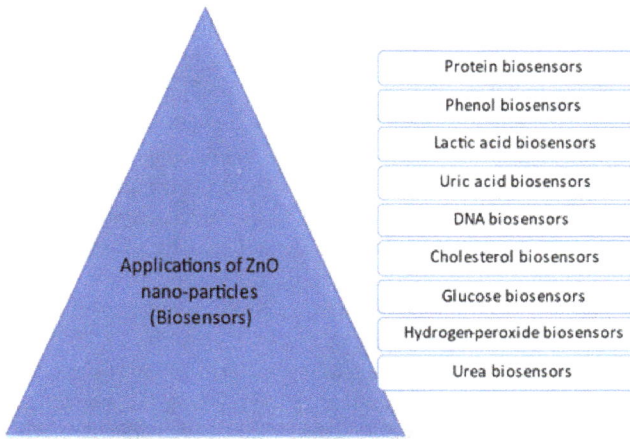

Protein biosensors

Phenol biosensors

Lactic acid biosensors

Uric acid biosensors

DNA biosensors

Cholesterol biosensors

Glucose biosensors

Hydrogen peroxide biosensors

Urea biosensors

Applications of ZnO nano-particles (Biosensors)

Figure 5: Applications of ZnO NPs as biosensors

9. Modern applications

9.1 Biosensor

Application of biosensors are wide spreading day by day. Estimation of cholesterol using chitosan are now consideringly important because it associated with heart disease, jaundice, diabetes and nephrosis and decrease level of cholesterol due anemia and hypothyroidism. Cholesterol estimation in different stages done by biosensing technology [50].Biosensors are the detecting agents that are based on biological materials. They are widely used in analysis such as biological analysis and food analysis. Zinc oxide are used in many types of bio sensors (Figure 5). It includes glucose biosensors, phenol biosensors, hydrogen peroxidase biosensor, cholesterol biosensors, urea biosensor etc.

9.2 MnFe₂O₄@SiO₂@PMIDA magnetic nanoparticles for antibody immobilization

In the past few years, some unexpected development takes place for the synthesis of probes which play a role for the acknowledgement of molecules specially biomolecules in several fields including medical science, biotechnology and chemistry. These probs used for study the biological functions. For environmental analysis, medical diagnostics and different chemical studies, enzyme-linked-immunosorbent used. It comprises precise antibody and fixes to an antigen. Antibody conjugated nanoparticles which provide enormous surface area in all the 3 dimensions in order to boost the loading capacity of proteins.so we speediness the number of immobilized substances for different diagnostic purposes and applications comprises therapeutic applications. In the field of immunoassay, the most reliable method for antibody immobilization is the only solution for the development of highly activated biosensors. This method is also helpful for the detection of trace amount targets. In this regard covalent immobilization is mostly used for stability point of view and for better yield [51]. On magnetic nanoparticles high surface density of antibodies provides enormous binding sites for antigen. Magnetic ferrite NPs are particularly used for their rich crystal chemistry which can be changed by tunning the magnetic field. These particles contain a cubic face centered structure along with tetrahedral and octahedral interstitial sites occupied by metal cation. The metal cations are mostly divalent and trivalent [52]. Synthesis of $MnFe_2O_4@SiO_2@PMIDA$ magnetic nanoparticles for anti-prostate specific membrane antigen (PSMA) immobilization. *Co precipitation* method is used for the synthesis of manganese ferrite nanoparticles and then coated by SiO_2 shell. Co-precipitation of $MnCl_2$ and $FeCl_3$ in water is done by presence of NAOH. It is only a unique method for preparation manganese ferrite nanoparticles. After its preparation it is than coated with silica. It is done by the hydrolysis of tetraethyl orthosilicate on the manganese ferrite nanoparticles.it is than treated with 0.1M HCL. After that for preparation of $MnFe_2O_4@SiO_2@PMIDA$, silica coated nanoparticle through ultrasonic probe disperses in water solution. The coupling agent that is phosphonomethyl iminodiacetic acid (PMIDA). After continuously stirring finally we got our product. The synthesis process is shown in Figure 6.

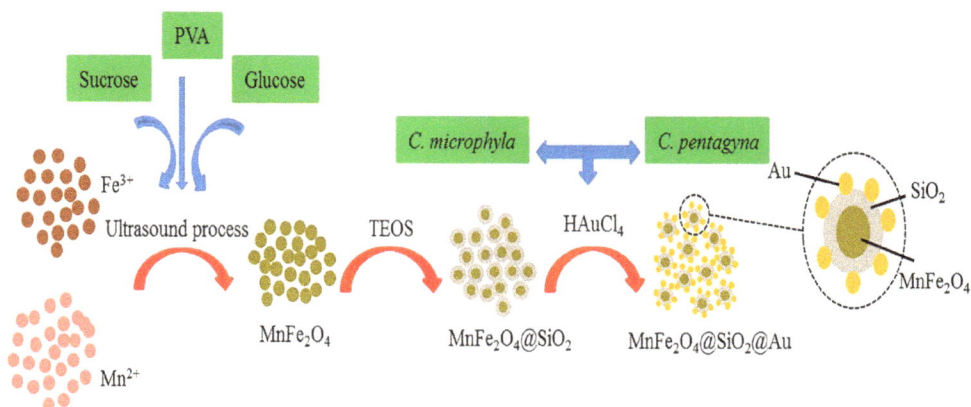

Figure 6: Steps for preparation of MnFe₂O₄@SiO₂@PMIDA

The silica coating became fabricated by way of postponing the magnetic nanoparticle in an ethanol-water and mixing with TEOS to shape middle-shell systems ($MnFe_2O4@SiO_2$). This step has been achieved so that it will stopover oxidation, agglomeration and, the extreme density of -OH on the silica surface allows in addition to change the particles through different steps. It is now functionalized.

Conclusion

The whole discussion in different sections indicates the importance of blooming area of research which need to explore in real sense in clinical, environmental, agriculture and industrial zones.

Acknowledgment

The authors are thankful to Higher Education Commission (HEC), Islamabad and Government College University, Faisalabad.

Materials Research Forum LLC
https://doi.org/10.21741/9781644901977-8

References

[1] O.M. Darwesh, S.S. Ali, I.A. Matter, T. Elsamahy, Y.A. Mahmoud, Enzymes immobilization onto magnetic nanoparticles to improve industrial and environmental applications, Methods Enzymol. 630 (2020) 481-502. https://doi.org/10.1016/bs.mie.2019.11.006

[2] Z. Ashkan, R. Hemmati, A. Homaei, A. Dinari, M. Jamlidoost, A. Tashakor, Immobilization of enzymes on nanoinorganic support materials: An update, Int J Biol Macromol. 168 (2021) 708-721. https://doi.org/10.1016/j.ijbiomac.2020.11.127

[3] S.A.R. Naqvi, K. Drlica, Fluoroquinolones as imaging agents for bacterial infection, Dalton Trans. 46 (2017) 14452-14460. https://doi.org/10.1039/C7DT01189J

[4] M.J. Cooney, Kinetic measurements for enzyme immobilization, Methods Mol Biol. 679 (2011) 207-225. https://doi.org/10.1007/978-1-60761-895-9_17

[5] S. Keller, S.P. Teora, G.X. Hu, M. Nijemeisland, D.A. Wilson, High-Throughput Design of Biocompatible Enzyme-Based Hydrogel Microparticles with Autonomous Movement, Angewandte Chemie. 57 (2018) 9814-9817. https://doi.org/10.1002/anie.201805661

[6] D.-M. Liu, J. Chen, Y.-P. Shi, Advances on methods and easy separated support materials for enzymes immobilization, TrAC Trends in Analytical Chemistry. 102 (2018) 332-342. https://doi.org/10.1016/j.trac.2018.03.011

[7] S.M.A. Shah, S.A.R. Naqvi, N. Munir, S. Zafar, M. Akram, J. Nisar, Antihypertensive and Antihyperlipidemic Activity of Aqueous Methanolic Extract of Rauwolfia Serpentina in Albino Rats, Dose-Response. 18 (2020) 1559325820942077. https://doi.org/10.1177/1559325820942077

[8] H. Vaghari, H. Jafarizadeh-Malmiri, M. Mohammadlou, A. Berenjian, N. Anarjan, N. Jafari, S. Nasiri, Application of magnetic nanoparticles in smart enzyme immobilization, Biotechnol Lett. 38 (2016) 223-233. https://doi.org/10.1007/s10529-015-1977-z

[9] T.A. Sherazi, T. Rehman, S.A.R. Naqvi, A.J. Shaikh, S.A. Shahzad, G. Abbas, R. Raza, A. Waseem, Surface functionalization of solid state ultra-high molecular weight polyethylene through chemical grafting, Applied Surface Science. 359 (2015) 593-601. https://doi.org/10.1016/j.apsusc.2015.10.080

[10] T.A. Sherazi, S. Zahoor, R. Raza, A.J. Shaikh, S.A.R. Naqvi, G. Abbas, Y. Khan, S. Li, Guanidine functionalized radiation induced grafted anion-exchange membranes for

solid alkaline fuel cells, International Journal of Hydrogen Energy. 40 (2015) 786-796. https://doi.org/10.1016/j.ijhydene.2014.08.086

[11] A. Rodriguez-Abetxuko, D. Sánchez-deAlcázar, P. Muñumer, A. Beloqui, Tunable Polymeric Scaffolds for Enzyme Immobilization, Frontiers in Bioengineering and Biotechnology. 8 (2020). https://doi.org/10.3389/fbioe.2020.00830

[12] U.T. Bornscheuer, Immobilizing Enzymes: How to Create More Suitable Biocatalysts, Angewandte Chemie International Edition. 42 (2003) 3336-3337. https://doi.org/10.1002/anie.200301664

[13] A. Arsalan, H. Younus, Enzymes and nanoparticles: Modulation of enzymatic activity via nanoparticles, Int J Biol Macromol. 118 (2018) 1833-1847. https://doi.org/10.1016/j.ijbiomac.2018.07.030

[14] T. Sherazi, R. Ullah, S. Naqvi, M.A. Rasheed, G. Ali, A. Shah, Y. Khan, Electrodeposition-assisted formation of anodized TiO2-CuO heterojunctions for solar water splitting, Applied Nanoscience. 11 (2020) 1-12. https://doi.org/10.1007/s13204-020-01557-x

[15] S.A. Shahzad, A. Sarfraz, M. Yar, Z.A. Khan, S.A.R. Naqvi, S. Naz, N.A. Khan, U. Farooq, R. Batool, M. Ali, Synthesis, evaluation of thymidine phosphorylase and angiogenic inhibitory potential of ciprofloxacin analogues: Repositioning of ciprofloxacin from antibiotic to future anticancer drugs, Bioorganic chemistry. 100 (2020) 103876. https://doi.org/10.1016/j.bioorg.2020.103876

[16] M. Sattar, H. Anwar, M. Faisal, G. Hussain, S. Irfan, A. Rasul, I. Mukhtar, M.U. Sohail, H. Muzaffar, A. Shaukat, S. Naqvi, S. Naqvi, Synergetic effects of GOS and Cu +2 nanoparticles as prebiotics on biochemical and metabolic hormonal profile in alloxan induced diabetic rats model, Pakistan journal of pharmaceutical sciences. 33 (2020) 1297-1302.

[17] M. Jamal, M. Khosa, M. Rashad, A. Mansha, S. Naqvi, Volumetric and Acoustic Behavior of Sodium Cyclamate in Aqueous System from 293.15 K to 318.15 K, Journal of Solution Chemistry. 45 (2016). https://doi.org/10.1007/s10953-016-0488-4

[18] L. Carneiro, R.J. Ward, Functionalization of paramagnetic nanoparticles for protein immobilization and purification, Anal Biochem. 540-541 (2018) 45-51. https://doi.org/10.1016/j.ab.2017.11.005

[19] M. Bilal, Y. Zhao, T. Rasheed, H.M.N. Iqbal, Magnetic nanoparticles as versatile carriers for enzymes immobilization: A review, Int J Biol Macromol. 120 (2018) 2530-2544. https://doi.org/10.1016/j.ijbiomac.2018.09.025

[20] M.F. Tahir, S.A. Bukhari, F. Anjum, M. Qasim, H. Anwar, S.A.R. Naqvi, Purification and modification of Cordia myxa gum to enhance its nutraceutical attribute as binding agent, Pakistan journal of pharmaceutical sciences. 32 (2019) 2245-2250.

[21] D.-M. Liu, J. Chen, Y.-P. Shi, Advances on methods and easy separated support materials for enzymes immobilization, J TrAC Trends in Analytical Chemistry. 102 (2018) 332-342. https://doi.org/10.1016/j.trac.2018.03.011

[22] L. Zhang, P. Wang, C. Wang, Y. Wu, X. Feng, H. Huang, L. Ren, B.-F. Liu, S. Gao, X.J.S.r. Liu, Bacteriophage T4 capsid as a nanocarrier for Peptide-N-Glycosidase F immobilization through self-assembly, 9 (2019) 1-13. https://doi.org/10.1038/s41598-019-41378-9

[23] W. Tischer, V. Kasche, Immobilized enzymes: crystals or carriers?, Trends Biotechnol. 17 (1999) 326-335. https://doi.org/10.1016/S0167-7799(99)01322-0

[24] A. Bari, Z.A. Khan, S.A. Shahzad, S.A. Raza Naqvi, S.A. Khan, H. Amjad, A. Iqbal, M. Yar, Design and syntheses of 7-nitro-2-aryl-4H-benzo[d][1,3]oxazin-4-ones as potent anticancer and antioxidant agents, Journal of Molecular Structure. 1214 (2020) 128252. https://doi.org/10.1016/j.molstruc.2020.128252

[25] E.D. Yushkova, E.A. Nazarova, A.V. Matyuhina, A.O. Noskova, D.O. Shavronskaya, V.V. Vinogradov, N.N. Skvortsova, E.F. Krivoshapkina, Application of Immobilized Enzymes in Food Industry, J Agric Food Chem. 67 (2019) 11553-11567. https://doi.org/10.1021/acs.jafc.9b04385

[26] T. Akkas, A. Zakharyuta, A. Taralp, C.W. Ow-Yang, Cross-linked enzyme lyophilisates (CLELs) of urease: A new method to immobilize ureases, Enzyme Microb Technol. 132 (2020) 109390. https://doi.org/10.1016/j.enzmictec.2019.109390

[27] M. Sharifi, A.Y. Karim, N. Mustafa Qadir Nanakali, A. Salihi, F.M. Aziz, J. Hong, R.H. Khan, A.A. Saboury, A. Hasan, O.K. Abou-Zied, M. Falahati, Strategies of enzyme immobilization on nanomatrix supports and their intracellular delivery, J Biomol Struct Dyn. 38 (2020) 2746-2762. https://doi.org/10.1080/07391102.2019.1643787

[28] M. Hoarau, S. Badieyan, E.N.G. Marsh, Immobilized enzymes: understanding enzyme - surface interactions at the molecular level, Org Biomol Chem. 15 (2017) 9539-9551. https://doi.org/10.1039/C7OB01880K

[29] S.A. Naqvi, J.K. Sosabowski, S.A. Nagra, M.M. Ishfaq, S.J. Mather, T. Matzow, Radiopeptide internalisation and externalization assays: cell viability and radioligand

integrity, Applied radiation and isotopes : including data, instrumentation and methods for use in agriculture, industry and medicine. 69 (2011) 68-74. https://doi.org/10.1016/j.apradiso.2010.09.005

[30] R.A. Wahab, N. Elias, F. Abdullah, S.K. Ghoshal, On the taught new tricks of enzymes immobilization: An all-inclusive overview, J Reactive Functional Polymers. 152 (2020) 104613. https://doi.org/10.1016/j.reactfunctpolym.2020.104613

[31] T. Jesionowski, J. Zdarta, B. Krajewska, Enzyme immobilization by adsorption: a review, Adsorption. 20 (2014) 801-821. https://doi.org/10.1007/s10450-014-9623-y

[32] S.A. Bukhari, N. Farah, G. Mustafa, S. Mahmood, S.A.R. Naqvi, Magneto-Priming Improved Nutraceutical Potential and Antimicrobial Activity of Momordica charantia L. Without Affecting Nutritive Value, Applied Biochemistry and Biotechnology. 188 (2019) 878-892. https://doi.org/10.1007/s12010-019-02955-w

[33] S.A. Shahzad, M. Yar, M. Bajda, L. Shahzadi, Z.A. Khan, S.A.R. Naqvi, S. Mutahir, N. Mahmood, K.M. Khan, Synthesis, thymidine phosphorylase inhibition and molecular modeling studies of 1,3,4-oxadiazole-2-thione derivatives, Bioorganic Chemistry. 60 (2015) 37-41. https://doi.org/10.1016/j.bioorg.2015.04.003

[34] M. Yaseen, Z. Farooq, M.H.R. Mahmood, S.A. Ahmad, S. Nazir, K.M. Anjum, S.A.R. Naqvi, Synthesis of Novel Symmetric Porphyrin Schiff Base Dimers by Solid-Liquid Reaction Methodology, Journal of Heterocyclic Chemistry. 56 (2019) 1520-1529. https://doi.org/10.1002/jhet.3526

[35] M. Bilal, S. Mehmood, T. Rasheed, H.M.N. Iqbal, Bio-Catalysis and Biomedical Perspectives of Magnetic Nanoparticles as Versatile Carriers, 5 (2019) 42. https://doi.org/10.3390/magnetochemistry5030042

[36] S.A. Naqvi, T. Matzow, C. Finucane, S.A. Nagra, M.M. Ishfaq, S.J. Mather, J. Sosabowski, Insertion of a lysosomal enzyme cleavage site into the sequence of a radiolabeled neuropeptide influences cell trafficking in vitro and in vivo, Cancer biotherapy & radiopharmaceuticals. 25 (2010) 89-95. https://doi.org/10.1089/cbr.2009.0666

[37] Z. Khan, S. Shahzad, A. Anjum, A. Bale, S. Naqvi, Synthetic approaches toward the reserpine, Synthetic Communications. 48 (2018) 1-20. https://doi.org/10.1080/00397911.2018.1434546

[38] H. Orhan, D. Aktaş Uygun, Immobilization of L-Asparaginase on Magnetic Nanoparticles for Cancer Treatment, Appl Biochem Biotechnol. 191 (2020) 1432-1443. https://doi.org/10.1007/s12010-020-03276-z

[39] M. Chang, Y.J. Chang, P.Y. Chao, Q. Yu, Exosome purification based on PEG-coated Fe3O4 nanoparticles, PLoS One. 13 (2018) e0199438. https://doi.org/10.1371/journal.pone.0199438

[40] M. Basit, M.H. Rasool, S.A.R. Naqvi, M. Waseem, B. Aslam, Biosurfactants production potential of native strains of Bacillus cereus and their antimicrobial, cytotoxic and antioxidant activities, Pakistan journal of pharmaceutical sciences. 31 (2018) 251-256.

[41] A. Ulu, S.A.A. Noma, S. Koytepe, B. Ates, Chloro-Modified Magnetic Fe(3)O(4)@MCM-41 Core-Shell Nanoparticles for L-Asparaginase Immobilization with Improved Catalytic Activity, Reusability, and Storage Stability, Appl Biochem Biotechnol. 187 (2019) 938-956. https://doi.org/10.1007/s12010-018-2853-9

[42] R. Dhankhar, V. Gupta, S. Kumar, R.K. Kapoor, P. Gulati, Microbial enzymes for deprivation of amino acid metabolism in malignant cells: biological strategy for cancer treatment, Appl Microbiol Biotechnol. 104 (2020) 2857-2869. https://doi.org/10.1007/s00253-020-10432-2

[43] M. Yar, L.R. Sidra, E. Pontiki, N. Mushtaq, R. Nasar, I. Khan, N. Mahmood, S. Naqvi, Z. Khan, S. Shahzad, Synthesis, in vitro lipoxygenase inhibition, docking study and thermal stability analyses of novel indole derivatives, Journal of the Iranian Chemical Society. 11 (2014). https://doi.org/10.1007/s13738-013-0308-3

[44] M. Yar, M. Bajda, R.A. Mehmood, L.R. Sidra, N. Ullah, L. Shahzadi, M. Ashraf, T. Ismail, S.A. Shahzad, Z.A. Khan, S.A. Naqvi, N. Mahmood, Design and Synthesis of New Dual Binding Site Cholinesterase Inhibitors: in vitro Inhibition Studies with in silico Docking, Letters in drug design & discovery. 11 (2014) 331-338. https://doi.org/10.2174/15701808113106660078

[45] M. Asif, S.A.R. Naqvi, T.A. Sherazi, M. Ahmad, A.F. Zahoor, S.A. Shahzad, Z. Hussain, H. Mahmood, N. Mahmood, Antioxidant, antibacterial and antiproliferative activities of pumpkin (cucurbit) peel and puree extracts - an in vitro study, Pakistan journal of pharmaceutical sciences. 30 (2017) 1327-1334.

[46] A. Basso, S. Serban, Industrial applications of immobilized enzymes-A review, Molecular Catalysis. 479 (2019) 110607. https://doi.org/10.1016/j.mcat.2019.110607

[47] M.L. Verma, S. Kumar, A. Das, J.S. Randhawa, M.J.E.C.L. Chamundeeswari, Chitin and chitosan-based support materials for enzyme immobilization and biotechnological applications, 18 (2020) 315-323. https://doi.org/10.1007/s10311-019-00942-5

[48] S. Singh, Zinc oxide nanoparticles impacts: cytotoxicity, genotoxicity, developmental toxicity, and neurotoxicity, Toxicol Mech Methods. 29 (2019) 300-311. https://doi.org/10.1080/15376516.2018.1553221

[49] P.K. Mishra, H. Mishra, A. Ekielski, S. Talegaonkar, B. Vaidya, Zinc oxide nanoparticles: a promising nanomaterial for biomedical applications, Drug Discov Today. 22 (2017) 1825-1834. https://doi.org/10.1016/j.drudis.2017.08.006

[50] R. Khan, A. Kaushik, P.R. Solanki, A.A. Ansari, M.K. Pandey, B.D. Malhotra, Zinc oxide nanoparticles-chitosan composite film for cholesterol biosensor, Anal Chim Acta. 616 (2008) 207-213. https://doi.org/10.1016/j.aca.2008.04.010

[51] M. Najafi, S. Bukhari, S. Nagra, S. Naqvi, The Substituent and Solvent Effects on the Antioxidant Activity of the Ferulic Acid Derivations, Journal of the Chemical Society of Pakistan. 36 (2014) 268-276.

[52] B. Aslibeiki, P. Kameli, H. Salamati, G. Concas, M. Salvador Fernandez, A. Talone, G. Muscas, D. Peddis, Co-doped MnFe(2)O(4) nanoparticles: magnetic anisotropy and interparticle interactions, Beilstein J Nanotechnol. 10 (2019) 856-865. https://doi.org/10.3762/bjnano.10.86

Chapter 9

Use of Nanomaterials-based Enzymes in Vaccine Production and Immunization

Mamata Singh[1], Paulin Nzeyimana[2], Ahumuza Benjamin[2], Amita Soumya[3],
N.P. Singh[4], Vivek Mishra[5,*]

[1]Department of Chemistry, GITAM School of Sciences, Gandhi Institute of Technology and Management (Deemed to be University), Bangalore, India

[2]Department of Mechanical Engineering, GITAM Institute of Technology, Gandhi Institute of Technology and Management (Deemed to be University), Bangalore, India

[3]Department of Chemistry, Presidency University, Bangalore, Karnataka

[4]Centre for NanoScience and Engineering (CeNSE), Indian Institute of Science, Bangalore, Karnataka, India

[5]Amity Institute of Click Chemistry Research and Studies (AICCRS), Amity University Uttar Pradesh, India

* vmishra@amity.edu

Abstract

The production of standard vaccines is increasing rapidly. The improvement is needed due to concerns of low immunogenicity, instability, and the need for more vaccines. To overcome these concerns, development of vaccines has been integrated with and facilitated by nanotechnology. Nanotechnology is increasingly playing a key role in vaccination by the development of NP-based delivery systems which have aided in increasing cellular and humoral immune responses. The nano carrier-based system facilitates the delivery of vaccine antigens to target cells and increases antigen resistance and immunogenicity. Many nano-sized particles have been studied and are being used as adjuvants and vehicles to deliver vaccine antigens. The efficiency of NPs as nanocarriers is due to their size and promoting specialized and selective immune responses. This chapter will focus on nanonzyme and their use in vaccine production and immunization.

Keywords

Nano-Enzyme, Vaccine Delivery, Immunity, Immune Response, Nanovaccinology

Contents

1.   Introduction

Synthetic enzymes have become of great interest to many scientists because of the many benefits and properties they have over natural enzymes [1–3]. Many synthetic materials, such as cyclodextrins, metal structures, organ selenium, and porphyrins [1,4,5], have been extensively studied and used to design and synthesize these enzymes in a variety of ways. Today, much research is being done in a class of synthetic enzymes called nano enzymes [6–8] that are nanomaterials with enzyme-like properties. These structures have features such as low cost, longevity, mass production, tunable catalytic activities, robust high performance compared to natural enzymes [2,9,10]. The advancement of biology and nanotechnology has provided new possibilities for formation of various nanozymes until many biological or functional components can be successfully integrated into a single nano-scaled system [11–13]. In addition, it is preferable that the enzyme produced not only mimics replicating the natural features of the enzymes but can also reflect other novel properties in biological applications. Blending of nanomaterials into biomimetic enzymes provides a simpler and better way of modulating the activity and durability of catalysts. It is good to note that the imitations by nanozymes tend to have low catalytic performances at times [6] hence cannot mimic the high stimulant activity of natural enzymes completely.

Vaccination helps in reducing a number of infectious diseases [14–16] and many vaccines have been licensed for use in combating various diseases. The first generation of vaccines is still very effective though not safe enough. There is a lot of effort being rendered to develop their suitable alternatives. The main drawbacks of DNA vaccines are due to oncogene activation and elicitation of anti-DNA antibodies and their low immunogenicity. Commercial drugs made from inactivated toxins require complex substances in their culture medium, which makes them inefficient and expensive to produce. Ordinary vaccines based on bone marrow transplants show the risk of mutation into pathogenic virulence, while inactive pathogen injections often lead to a weak immune response. Chitosan nanoparticles (CSNPs) and Mannosylated Chitosan Nanoparticles (MCH NPs) are loaded with recurrent hepatitis B surface antigen (rHBsAg) as the adjuvants and show a progressive release pathway, with no toxic in-vivo effects and very high antibodies[17–19]. In addition, nanoparticles containing high levels of fat elicited; a strong immune response compared to live virus vaccines.

2. Enzymes

Enzymes are soluble, colloidal, and organic catalysts generated by living cells. They stimulate chemical reactions by reducing activation capacity/energy and, in the process, remain unchanged [20–22]. Their main function is to stimulate the biochemical reaction of organisms and they remain unchanged at the end of the reaction [23,24]. Enzymes can be classified as natural or synthetic. The synthetic enzymes are subsequently modified and manipulation by chemical route. They are able to perform the function like former natural enzymes while natural enzymes are present in plants and animals [20,22]. Despite their different conditions, both play the same role in stimulating bio-chemical reactions.

2.1 How enzymes work

Enzymes stimulate chemical reactions by binding to specific substrates. The specification is achieved by binding packets with consistent charge, and hydrophilic/hydrophobic markers on substrates which is why enzymes can differentiate between identical substrate molecules to select chemo, regioselective and stereospecific [23,24]. The process of reactivating a chemical reaction can be explained by two models, namely – 1) *Lock and key model and 2) Induced fit model* [24,25].

In the lock and key model (Fig. 1), the enzyme and the substrate have certain corresponding geometric elements that directly co-relate to each other in terms of geometry to perform biochemical reaction. While an *induced fit model* (Fig. 2) the active site of enzymes slightly modified itself with respect to substrate to interact with it and perform biochemical activity.

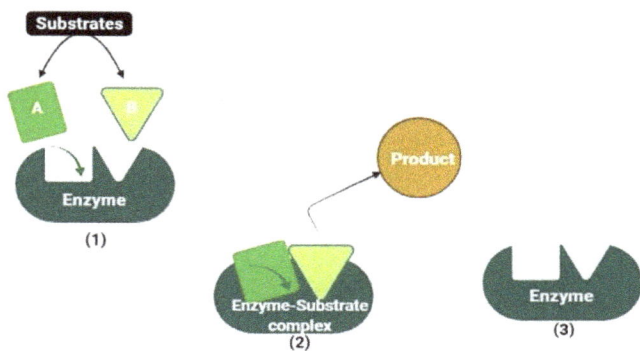

Figure 1. Key-lock model- functions of the enzyme (self redrawn)

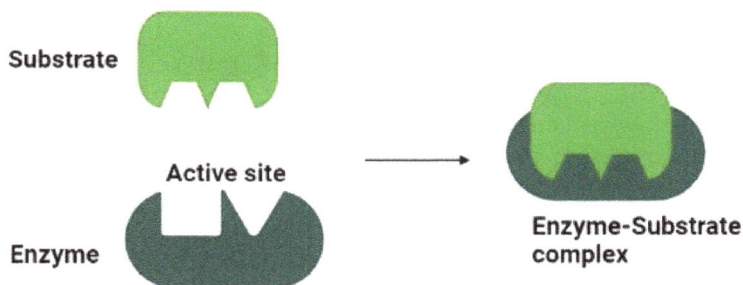

Figure 2. Induced fit model-functions of the enzyme (self redrawn)

2.2   Natural and Artificial Enzymes

• *Natural enzymes*

Natural enzymes are highly efficient and versatile biocatalysts that occur naturally in plants and animals [26–28].

• *Artificial Enzymes*

An artificial enzyme is a synthetic, organic molecule or ion that performs a specific natural enzyme activity 29–32].

Table 1: Natural and Artificial classification of enzymes

Natural enzymes	Artificial enzymes
Protease	Cyclodextrins
Phospho Kinases	XNAzymes
Oxygenases	Amylase
EcoRI	Lysozyme
Trypsin	Urease
Cellulose	Amyloglucosidase

3.   Nanozymes

Nanozymes are class of synthetic enzymes with nanoscale size that exhibit enzyme-like characteristics [33–35]. Recently, researchers have discovered that certain substances can act as synthetic enzymes that include fullerenes, cyclodextrins [36], polymerspl [37,38], dendrimers, porphyrins, metal complexes [39,40], and other biomolecules. Fe_3O_4 nanoparticles [41] as peroxidase have been found to have many properties that can make them act as an enzyme—Nanozymes are divided into two categories.

- *Nanomaterial hybrid enzymes-* This enzymatic catalytic group achieves improved stability and durability by modifying it with the help of nanomaterials [42].

- *Nanomaterials that have natural enzymatic catalytic pro*perties that show a similar mechanism as an enzyme that catalyzes similar biocatalytic reactions.

Nanomaterial synthetic enzymes have both chemical and biological properties[6,43]. They have advantages over natural enzymes, such as very low cost, high durability, and long-lasting properties. They can also catalyze artificial bioprocesses such as biorthogonal catalysis. Depending on their properties, nanozymes are used in biosensing, environmental therapy, diagnosis and treatment of diseases, antibacterial agents, and cytoprotecting against cellular biomolecules, etc.

Table 2: Examples of Nanozymes

Enzyme	Example	Application
Oxidase	MSN-Au	Antibacterial
	Au@Pt	Immunoassay
	$PtCo@MnO_2$	Cancer therapy
Peroxidase	PtCu	Phenol degradation
	CuS	Immunodetection
Hydrolase	DMAE	antibacterial
	VE CeO_2	Degradation nerve agents
	AuNPs@POMD-8pe	AD therapy

Like natural enzymes, nanozymes are affected by some factors such as pH in the surrounding environment, surface morphology, optimum temperature, and ions in reaction and light that can affect reactions by blocking the interaction between the enzyme and the substrate.

4. Nanozymes in vaccine production and immunization

Nanozymes are a kind of nanomaterials with enzymatic catalytic properties. Nanozymes have a host of various interesting features such as low cost and high durability that have served as a key point of interest for their applications in the medical and biological fields [6,33,43–45]. Since, in their discovery, nanozymes have been very useful in producing many vaccines [46–48]. A vaccine is a biological preparation that provides acquired immunity to a particular infectious disease [14,15,49]. Today, one of the major roles played by these nanozymes can be traced to the production of COVID-19 vaccine [7].

They ruled as a mimicking enzyme and play a key role in the body's immune response [50,51]. Immunity is the body's ability to prevent infections [1,52]. There are many types of immunity, namely, natural immunity, innate immunity, adaptive immunity, and acquired immunity [1,53,54].

Nanozyme vaccines are also ideal in cases where resources are not sufficient, or the population density is high due to their autonomous characteristics. In this regard, one can use single dose infusions, nanofilm based vaccines, and microneedle patches.

Genetic nano vaccines have also played an influential role in the treatment of Ebola, HIV-1, and oral infections [55,56]. DNA and RNA vaccines are preferred to traditional vaccines owing to their simple purification and low production cost, however, their targeted delivery faces some constraints due to their specificity and stability [14,57–60]. The inefficiency in traditional delivery systems has invoked nanotechnology to facilitate the delivery through nanomedicine which is more efficacious [61–64]. A combination of RNA and nanocarriers can be used to treat serious illnesses, autoimmune diseases, and neurological disorders by introducing a small amount of disruptive RNA (siRNA), including mRNA-1273 vaccine, built nanocarrier system [64]. Nanocarriers that deliver antigens prevent their premature deterioration and help to convert them into active immunogens hence reducing side effects. Yersinia pestis F1-antigen - coated AuuPs [65] and Influenza antigen H1N1 - conjugated chitosan NPs, for example, improved responses to cytokine and antibody levels compare to mice controlled by unintended antigens. It is shown to be due to the stability of the vaccine antigen and better immunity due to the antigens associated with NPs [34,46,66]. Vaccine delivery by nanoparticle (NPb-Vs) programs is done by-

- Embedding the antigen in the nanocarrier which presents the RNA or DNA directly to the APCs that translate them on the surface of the cell into the corresponding antigens [67–70]. In this sense, RNA is preferred since it can be converted directly into cell cytoplasm, unlike DNA, which is only translated after reaching the target cell. This method also prevents the antigens from proteolytic degradation and can also work with a local cache effect, which increases the antigen exposure time in the immune system.

- By attaching antigens to the Nano network area to ensure exposure to the environment.

4.1 Nanomaterial-based enzymes in vaccine production

Nano enzymatic materials and nanoparticles have played a major role in producing various vaccines such as Nanoflu, Nuvec, Novavax, COVID-19 vaccine, etc.

4.1.1 Nanoflu

NanoFlu contains Novavax's saponin-based Matrix-M adjuvant, which is powerful, well-tolerated, and stimulates both high and strong responses such as CD4 and CD8 T-cell responses [71]. Recombinant flu vaccines have an important advantage. Owing to this, once a commercial license has been issued, large amounts of vaccines can be produced at a low cost, without the use of live flu viruses or eggs. To develop the vaccine, researchers combined HA proteins with protein-forming nanoparticles which indicate the HA proteins

the immune system has to respond. The team developed four types of nanoparticles, each using HA from a different type of flu and the "mosaic" nanoparticles incorporating all four HAs into each nanoparticle. The nanoparticle vaccine elevated antibody responses against flu strains that were better than those elicited by the commercial drugs and also provided close immunity to the flu that the commercial drug did not protect against. The nanoparticle vaccine includes only two of the 18 influenza A virus HA subtypes, H1 and H3. But the cocktail vaccine of four nanoparticles provided 73% protection against viruses containing H5 and H7 subtypes, while the mosaic vaccine provided 92% protection and the commercial drug provided 12% protection only.

4.1.2 COVID-19 vaccine

Nanomedicine has played a role in the formulation, delivery, and management of the COVID-19 vaccine [72–76]. Vaccination is considered to be the most promising anti-coronavirus strategy since the vaccines produced to combat this virus possess a strong antigen presentation and good antigen enhancement. Nanoparticles are used as carriers of vaccines in the targeted delivery of antigens and can be converted into inhibitors to enhance the immune response [74,77,78]. Examples include the mRNA transported by the liposomo nanoparticle [79–81] to the cytoplasm of trapped cells, where it can be synthesized into antigen proteins to initiate the production of antibodies. Traditional vaccines and their derivatives retreat to pathogenic toxicity or low immunogenicity at times, but this can be addressed by using and modifying NPs. Nanoparticles with properties such as rapid penetration of mucous can cross cellular boundaries and act with fewer toxins in the lungs, thus acting in the treatment of respiratory diseases. Syncytial virus infection can be treated by using NPs such as F -VLP protein based virus which regenerate parts of the body cells such as IFN-γ, and TNF-α. Modifying amino acids and using peptide inhibitors binds peptide ligands to the epitome of the B cell from the SARS-B HRC1 spike protein which helps in fighting the SARS-COV. The binding activity of an ACE2-isolated peptide inhibitor capable of blocking SARS-CoV-2 can be increased by binding to multiple nanocarrier-binding peptides. S-protein is used as the target for COVID-19 vaccines such as mRNA-1273 which target S CoVs protein.

How vaccines work

Vaccines help improve the immune system by mimicking an infection [82–84]. This type of mimicking by the Vaccine is harmless and doesn't trigger the immune system's production of antibodies and T-lymphocytes which is meant to fight the actual infection in the future [85]. Hence, body doesn't produce B-lymphocytes and T-lymphocytes immediately after vaccination. Therefore, it is possible for a vaccinated person to show symptoms if he was infected shortly before vaccination or to acquire the disease and show symptoms immediately after vaccination as the vaccine has not yet had sufficient time to create self-defense in body. In some cases, it is common to show minor symptoms, e.g., headache, cold, cough due to some stimulated infection after receiving the vaccine. This reaction is expected as the body builds up the immune system.

Function of the COVID-19 vaccine

- Spike protein type coronavirus is implicated in harmless chimpanzee virus
- The chimpanzee virus is genetically modified to prevent it from growing in humans
- Once there, cells study genetic information to produce millions of copies of the spike protein
- Fragments of protein stimulate the immune system to produce antibodies that can protect the body when the virus actually invades the body.

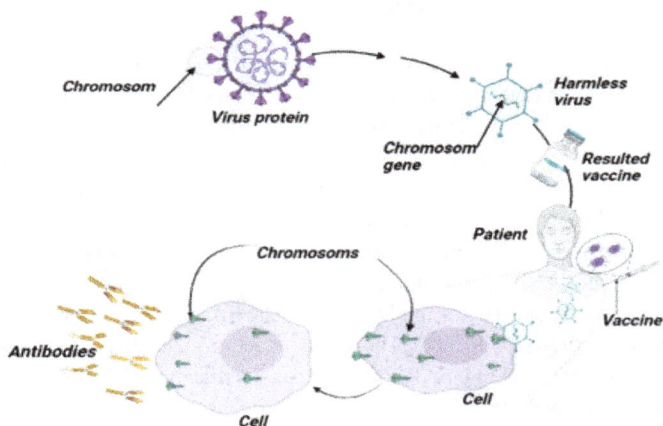

Fig. 3. Working of COVID-19 Vaccine (self redrawn)

4.2   Nanomaterial-based enzymes in immunization

The discovery, supplements have always been combined with vaccines to boost the body's immune response [86–88]. Many natural and industrially made chemicals have been identified as adjuvants, but aluminum-based chemicals remain the dominant additives in human vaccines [89]. The immune response caused by alum was believed to be a result of the depot effect in the injection site, which increases the antigen exposure in the immune system, thus getting a better response [89–91]. However, Holt challenged the depot effect theory by excising the injection site of a guinea pig which did not stop the humorous response from developing, thus concluding that the previous approach did not serve as the correct explanation of the immune response caused by alum. Current studies on alum adjuvants have observed that these adjuvants release antigens rapidly. Also, ~80% of $AlPO_4$ adsorbed tetanus toxoid [92–94] could not be traced at the injection site after 4 hrs of administering the Vaccine. Alum plays a significant role in detecting its electrostatic interactions with lipopolysaccharide and its desired effect on specific antigen proteins that lead to aluminum oxyhydroxide metal sheets. Aluminum particles usually have a diameter <10μm, and their antigens can easily be phagocytosed, unlike those without aluminum.

Particulate molecules produced in living organisms may lead to inflammasome, which activates Nlrp3 (cytoplasmic protein), causing eosinophilic infiltration and increased antigen and MHCII expression cell activity. The activation of Nlrp3 also stimulates the production of pro-inflammatory cytokines. One of these cytokines is IL-1β which produces T-cell-based antibodies in living tissues[95].

Alum facilitates antibodies andreates CD8 and can trigger cellular responses between CD4[96–98], creating cellular memory in vivo. The cellular memory created helps in producing long-term immune defenses and also protects the body against pathogens.

It is important to note that some toxic complications are associated with alum compounds that could threaten the body's health as they can cause diseases such as Alzheimer's disease and Lou Gehrig's disease. Hence safer adjuvants should be produced to avoid such scenarios.

The purpose of immunization is to ensure that the Vaccine administered into the human body mimics the natural immune system to respond to any infection consistently as the natural body does. When this is achieved, the body is said to have acquired an adaptive immune response. The key players here are the antigen-producing cells that use pattern recognition receptors to detect microorganisms. The microbial surface recognition process starts with the maturation of antigen-presenting cells. After that, MHC molecules are dispersed from one cell component to another, chemokines and cytokines are secreted, morphological changes and cytoskeletal reorganization occur. When surface recognition is done, antigen-specific antibodies are produced, and the T-cell memory is formed with the help of peptides produced by protein processing in the body that are found to be expressed in MHC II molecules as identified by CD4 + T cells. The body must first adapt to the

Nanomaterial-Supported Enzymes   Materials Research Forum LLC
Materials Research Foundations **126** (2022) 240-260   https://doi.org/10.21741/9781644901977-9

immunity induced by the vaccine before the vaccine can respond and deal with the pathogens [52].

The role of NPs in immunization is mainly to deliver antigens to their target locations. Recent studies show an increase in efficiency in dendritic cell acquisition and detection of antigen-presenting cells when NPs are used as antigen delivery vehicles. The same result is given by Chithrani et al. after a study on gold NPs using plasma-based spectroscopy where a significant increase in acquisition and detection was detected. This high detection rate of NPs by the cells also directly impacts the type of immune response produced.

The interaction between antigen-producing cells and particles is also affected by ground charge. The charged particles are easily transported to cells since the molecular cells are negatively charged in nature and particle formation where round-shaped cells are readily available, unlike the rod-shaped.

While PLGA microparticles themselves attach macrophages to phagocytosis, there are various ways NP can be inserted internally; some NPs can be taken up by HeLa cells through the clathrin-dependent endocytosis and others via the independent pathway of cholesterol, non-caveolar, and non-clathrin.

Poly (amino acids) NPs containing synthetic ovalbumin have the ability to respond to cellular immune responses. This is because of the induction of significantly higher IgG, IgG1, and IgG2a when CD4+ and CD8+ T cells are activated, producing IFNγ, which triggers the Ig phase IgG2a.

In the same way, a more resistant immune response to the marine model is observed when the core antigen of Hepatitis B [99–101] is added to PLGA NPs (300 nm).

When vaccination is done using PLA microparticles (2-8 nm), IL-4 is stimulated following the Th2 response. VLPs produce more robust immune responses against infections in both animal species and human clinical trials. VLP vaccines can introduce antigens to their natural state, such as proteins bound to the membrane and not the soluble ectodomain. This helps VLPs in improving the production of antibodies by mimicking the formation of a native virus. When VLPs are in vivo, an antigen depot is formed, allowing the antigens to be released, gradually increasing their exposure to the immune system and strengthening their dose. The release of the antigens is determined by the pharmacodynamics of each compound, whether slowly or at the depot. In general, particle formation reduces the need for adding more adjuvants to the vaccines, thus managing the price and a boasting effect due to the increased increase in particles by antigen cells.

References

[1] D.J. Irvine, M.A. Swartz, G.L. Szeto, Engineering synthetic vaccines using cues from natural immunity, Nature Materials. 12 (2013) 978-990. https://doi.org/10.1038/nmat3775

[2] Enzyme Technology: Application and Commercial Production of Enzymes, (n.d.). https://www.biologydiscussion.com/enzymes/enzyme-technology/enzyme-technology-application-and-commercial-production-of-enzymes/10185 (accessed July 16, 2021).

[3] J. Shah, S. Dave, A. Vyas, M. Shah, H. Arya, A. Gajipara, A. Vijapura, M. Bakshi, P. Thakore, R. Shah, V. Saxena, A. Shamal, S. Singh, Nanomaterials-Based Next Generation Synthetic Enzymes: Current Challenges and Future Opportunities in Biological Applications, Nanotechnology in Modern Animal Biotechnology: Concepts and Applications. (2019) 37-58. https://doi.org/10.1016/B978-0-12-818823-1.00004-1

[4] G. Ibrahim Fouad, A proposed insight into the antiviral potential of metallic nanoparticles against novel coronavirus disease-19 (COVID-19), Bulletin of the National Research Centre. 45 (2021). https://doi.org/10.1186/s42269-021-00487-0

[5] E.M. Melchor-Martínez, N.E. Torres Castillo, R. Macias-Garbett, S.L. Lucero-Saucedo, R. Parra-Saldívar, J.E. Sosa-Hernández, Modern World Applications for Nano-Bio Materials: Tissue Engineering and COVID-19, Frontiers in Bioengineering and Biotechnology. 9 (2021). https://doi.org/10.3389/fbioe.2021.597958

[6] Y. Huang, J. Ren, X. Qu, Nanozymes: Classification, Catalytic Mechanisms, Activity Regulation, and Applications, Chemical Reviews. 119 (2019) 4357-4412. https://doi.org/10.1021/acs.chemrev.8b00672

[7] M. Kumawat, A. Umapathi, E. Lichtfouse, H.K. Daima, Nanozymes to fight the COVID-19 and future pandemics., Environmental Chemistry Letters. (2021) 1-7. https://doi.org/10.1007/s10311-021-01252-5

[8] M. Kumawat, A. Umapathi, E. Lichtfouse, H.K. Daima, Nanozymes to fight the COVID-19 and future pandemics, Environmental Chemistry Letters. (2021). https://doi.org/10.1007/s10311-021-01252-5

[9] W. Tao, H.S. Gill, M2e-immobilized gold nanoparticles as influenza A vaccine: Role of soluble M2e and longevity of protection, Vaccine. 33 (2015) 2307-2315. https://doi.org/10.1016/j.vaccine.2015.03.063

[10] D.L. Jarvis, Developing baculovirus-insect cell expression systems for humanized recombinant glycoprotein production, Virology. 310 (2003) 1-7. https://doi.org/10.1016/S0042-6822(03)00120-X

[11] C. Weiss, M. Carriere, L. Fusco, L. Fusco, I. Capua, J.A. Regla-Nava, M. Pasquali, M. Pasquali, M. Pasquali, J.A. Scott, F. Vitale, F. Vitale, M.A. Unal, C. Mattevi, D. Bedognetti, A. Merkoçi, A. Merkoçi, E. Tasciotti, E. Tasciotti, A. Yilmazer, A. Yilmazer, Y. Gogotsi, F. Stellacci, L.G. Delogu, Toward Nanotechnology-Enabled Approaches against the COVID-19 Pandemic, ACS Nano. 14 (2020) 6383-6406. https://doi.org/10.1021/acsnano.0c03697

[12] M.G. Sharaf, S. Cetinel, L. Heckler, K. Damji, L. Unsworth, C. Montemagno, Nanotechnology-Based Approaches for Ophthalmology Applications: Therapeutic and Diagnostic Strategies., Asia-Pacific Journal of Ophthalmology (Philadelphia, Pa.). 3 (n.d.) 172-80. https://doi.org/10.1097/APO.0000000000000059

[13] I.Y. Wong, S.N. Bhatia, M. Toner, Nanotechnology: Emerging tools for biology and medicine, Genes and Development. 27 (2013) 2397-2408. https://doi.org/10.1101/gad.226837.113

[14] J.J. Donnelly, B. Wahren, M.A. Liu, DNA Vaccines: Progress and Challenges, The Journal of Immunology. 175 (2005) 633-639. https://doi.org/10.4049/jimmunol.175.2.633

[15] A. Facciolà, G. Visalli, P. Laganà, V. la Fauci, R. Squeri, G.F. Pellicanò, G. Nunnari, M. Trovato, A. di Pietro, The new era of vaccines: The "nanovaccinology," European Review for Medical and Pharmacological Sciences. 23 (2019) 7163-7182.

[16] D.A.G. Skibinski, B.C. Baudner, M. Singh, D.T. O'hagan, Combination vaccines, Journal of Global Infectious Diseases. 3 (2011) 63-72. https://doi.org/10.4103/0974-777X.77298

[17] N. Ketabchi, M. Naghibzadeh, M. Adabi, S.S. Esnaashari, R. Faridi-Majidi, Preparation and optimization of chitosan/polyethylene oxide nanofiber diameter using artificial neural networks, Neural Computing and Applications. 28 (2017) 3131-3143. https://doi.org/10.1007/s00521-016-2212-0

[18] D.R. Bhumkar, H.M. Joshi, M. Sastry, V.B. Pokharkar, Chitosan reduced gold nanoparticles as novel carriers for transmucosal delivery of insulin, Pharmaceutical Research. 24 (2007) 1415-1426. https://doi.org/10.1007/s11095-007-9257-9

[19] K. Zhao, G. Chen, X. ming Shi, T. ting Gao, W. Li, Y. Zhao, F. qiang Zhang, J. Wu, X. Cui, Y.F. Wang, Preparation and Efficacy of a Live Newcastle Disease Virus Vaccine Encapsulated in Chitosan Nanoparticles, PLoS ONE. 7 (2012). https://doi.org/10.1371/journal.pone.0053314

[20] WP (William P. Jencks, Catalysis in chemistry and enzymology, (1987) 836.

[21] K. Hult, P. Berglund, Engineered enzymes for improved organic synthesis, Current Opinion in Biotechnology. 14 (2003) 395-400. https://doi.org/10.1016/S0958-1669(03)00095-8

[22] R. Boyer, Chapter 6: Enzymes I, Reactions, Kinetics, and Inhibition, Concepts in Biochemistry. (2002) 137-8.

[23] D. Blow, So do we understand how enzymes work?, Structure. 8 (2000) R77-R81. https://doi.org/10.1016/S0969-2126(00)00125-8

[24] M.M. Cox, D.L. Nelson, Chapter 6.2: How enzymes work, Lehninger Principles of Biochemistry. (2013) 195. http://www.amazon.co.uk/Lehninger-Principles-Biochemistry-David-Nelson/dp/1464109621/ref=sr_1_1?s=books&ie=UTF8&qid=1425406097&sr=1-1&keywords=9781464109621 (accessed July 18, 2021).

[25] Suzuki H, S. H, Chapter 8: Control of Enzyme Activity, How Enzymes Work: From Structure to Function. (2015) 141-69. https://doi.org/10.1201/b18087-9

[26] R. Ravee, H.H. Goh, H.-H. Goh, discovery of digestive enzymes in carnivorous plants with focus on proteases, PeerJ. 6 (2018) e4914. https://doi.org/10.7717/peerj.4914

[27] P. Morino, F. Mascagni, A. McDonald, T. Hökfelt, Cholecystokinin corticostriatal pathway in the rat: Evidence for bilateral origin from medial prefrontal cortical areas, Neuroscience. 59 (1994) 939-52. https://doi.org/10.1016/0306-4522(94)90297-6

[28] Enzyme Definition and Classification - Creative Enzymes, (n.d.). https://www.creative-enzymes.com/resource/enzyme-definition-and-classification_18.html (accessed July 18, 2021).

[29] D.J. Mikolajczak, A.A. Berger, B. Koksch, Catalytically Active Peptide-Gold Nanoparticle Conjugates: Prospecting for Artificial Enzymes, Angewandte Chemie. 132 (2020) 8858-8867. https://doi.org/10.1002/ange.201908625

[30] Y. Lin, J. Ren, X. Qu, Catalytically Active Nanomaterials: A Promising Candidate for Artificial Enzymes, Accounts of Chemical Research. 47 (2014) 1097-1105. https://doi.org/10.1021/ar400250z

[31] Q. Wang, X. Zhang, L. Huang, Z. Zhang, S. Dong, GOx@ZIF-8(NiPd) Nanoflower: An Artificial Enzyme System for Tandem Catalysis, Angewandte Chemie International Edition. 56 (2017) 16082-16085. https://doi.org/10.1002/anie.201710418

[32] H.B. Albada, F. Soulimani, B.M. Weckhuysen, R.M.J. Liskamp, Scaffolded amino acids as a close structural mimic of type-3 copper binding sites, Chemical Communications. (2007) 4895-7. https://doi.org/10.1039/b709400k

[33] L. Gao, X. Yan, Nanozymes: an emerging field bridging nanotechnology and biology, Science China Life Sciences. 59 (2016) 400-402. https://doi.org/10.1007/s11427-016-5044-3

[34] L. Pasquato, P. Pengo, P. Scrimin, Nanozymes: Functional Nanoparticle-based Catalysts, Supramolecular Chemistry. 17 (2005) 163-171. https://doi.org/10.1080/10610270412331328817

[35] L. Huang, J. Chen, L. Gan, J. Wang, S. Dong, Single-atom nanozymes, Science Advances. 5 (2019) eaav5490. https://doi.org/10.1126/sciadv.aav5490

[36] Y. Zhao, Y. Huang, H. Zhu, Q. Zhu, Y. Xia, Three-in-One: Sensing, Self-Assembly, and Cascade Catalysis of Cyclodextrin Modified Gold Nanoparticles, Journal of the American Chemical Society. 138 (2016) 16645-16654. https://doi.org/10.1021/jacs.6b07590

[37] T. Merdan, J. Kopeček, T. Kissel, Prospects for cationic polymers in gene and oligonucleotide therapy against cancer, Advanced Drug Delivery Reviews. 54 (2002) 715-758. https://doi.org/10.1016/S0169-409X(02)00046-7

[38] N. Pippa, M. Gazouli, S. Pispas, Recent Advances and Future Perspectives in Polymer-Based Nanovaccines., Vaccines. 9 (2021). https://doi.org/10.3390/vaccines9060558

[39] H. Cheng, Y. Liu, Y. Hu, Y. Ding, S. Lin, W. Cao, Q. Wang, J. Wu, F. Muhammad, X. Zhao, D. Zhao, Z. Li, H. Xing, H. Wei, Monitoring of Heparin Activity in Live Rats Using Metal-Organic Framework Nanosheets as Peroxidase Mimics, Analytical Chemistry. 89 (2017) 11552-11559. https://doi.org/10.1021/acs.analchem.7b02895

[40] L. Qin, X. Wang, Y. Liu, H. Wei, 2D-Metal-Organic-Framework-Nanozyme Sensor Arrays for Probing Phosphates and Their Enzymatic Hydrolysis, Analytical Chemistry. 90 (2018) 9983-9989. https://doi.org/10.1021/acs.analchem.8b02428

[41] K. Pusic, Z. Aguilar, J. McLoughlin, S. Kobuch, H. Xu, M. Tsang, A. Wang, G. Hui, Iron oxide nanoparticles as a clinically acceptable delivery platform for a recombinant blood-stage human malaria vaccine, FASEB Journal. 27 (2013) 1153-1166. https://doi.org/10.1096/fj.12-218362

[42] J. Xi, G. Wei, L. An, Z. Xu, Z. Xu, L. Fan, L. Gao, Copper/Carbon Hybrid Nanozyme: Tuning Catalytic Activity by the Copper State for Antibacterial Therapy, Nano Letters. 19 (2019) 7645-7654. https://doi.org/10.1021/acs.nanolett.9b02242

[43] M. Liang, X. Yan, Nanozymes: From New Concepts, Mechanisms, and Standards to Applications, Accounts of Chemical Research. 52 (2019) 2190-2200. https://doi.org/10.1021/acs.accounts.9b00140

[44] Y. Huang, J. Ren, X. Qu, Nanozymes: Classification, Catalytic Mechanisms, Activity Regulation, and Applications, Chemical Reviews. 119 (2019) 4357-4412. https://doi.org/10.1021/acs.chemrev.8b00672

[45] J.J. Gooding, Can Nanozymes Have an Impact on Sensing?, ACS Sensors. 4 (2019) 2213-2214. https://doi.org/10.1021/acssensors.9b01760

[46] L.G.D.A.S.B.D.A. S Al-Halifa, Nanoparticle-based vaccines against respiratory viruses, Front Immunol. 10 (2019) 22. https://doi.org/10.3389/fimmu.2019.00022

[47] K.C. Petkar, S.M. Patil, S.S. Chavhan, K. Kaneko, K.K. Sawant, NK. Kunda, I.Y. Saleem, An Overview of Nanocarrier-Based Adjuvants for Vaccine Delivery., Pharmaceutics. 13 (2021). https://doi.org/10.3390/pharmaceutics13040455

[48] P. Li, Z. Luo, P. Liu, N. Gao, Y. Zhang, H. Pan, L. Liu, C. Wang, L. Cai, Y. Ma, Bioreducible alginate-poly(ethylenimine) nanogels as an antigen-delivery system robustly enhance vaccine-elicited humoral and cellular immune responses, Journal of Controlled Release. 168 (2013) 271-279. https://doi.org/10.1016/j.jconrel.2013.03.025

[49] I.P. Nascimento, L.C.C. Leite, Recombinant vaccines and the development of new vaccine strategies, Brazilian Journal of Medical and Biological Research. 45 (2012) 1102-1111. https://doi.org/10.1590/S0100-879X2012007500142

[50] Q. Wang, H. Wei, Z. Zhang, E. Wang, S. Dong, Nanozyme: An emerging alternative to natural enzyme for biosensing and immunoassay, TrAC Trends in Analytical Chemistry. 105 (2018) 218-224. https://doi.org/10.1016/j.trac.2018.05.012

[51] TAPF Pimentel, Z. Yan, S.A. Jeffers, K. v. Holmes, R.S. Hodges, P. Burkhard, Peptide nanoparticles as novel immunogens: Design and analysis of a prototypic severe acute respiratory syndrome vaccine, Chemical Biology and Drug Design. 73 (2009) 53-61. https://doi.org/10.1111/j.1747-0285.2008.00746.x

[52] B. Pulendran, R. Ahmed, Translating innate immunity into immunological memory: Implications for vaccine development, Cell. 124 (2006) 849-863. https://doi.org/10.1016/j.cell.2006.02.019

[53] T.H. Mogensen, Pathogen recognition and inflammatory signaling in innate immune defenses, Clinical Microbiology Reviews. 22 (2009) 240-273. https://doi.org/10.1128/CMR.00046-08

[54] Y. Honda-Okubo, F. Saade, N. Petrovsky, AdvaxTM, a polysaccharide adjuvant derived from delta inulin, provides improved influenza vaccine protection through broad-based enhancement of adaptive immune responses, Vaccine. 30 (2012) 5373-5381. https://doi.org/10.1016/j.vaccine.2012.06.021

[55] L. Xu, Y. Liu, Z. Chen, W. Li, Y. Liu, L. Wang, Y. Liu, X. Wu, Y. Ji, Y. Zhao, L. Ma, Y. Shao, C. Chen, Surface-engineered gold nanorods: Promising DNA vaccine adjuvant for HIV-1 treatment, Nano Letters. 12 (2012) 2003-2012. https://doi.org/10.1021/nl300027p

[56] C.N. Fries, J.-L. Chen, M.L. Dennis, N.L. Votaw, J. Eudailey, B.E. Watts, K.M. Hainline, D.W. Cain, R. Barfield, C. Chan, M.A. Moody, B.F. Haynes, K.O. Saunders, S.R. Permar, G.G. Fouda, J.H. Collier, HIV envelope antigen valency on peptide nanofibers modulates antibody magnitude and binding breadth, Scientific Reports. 11 (2021) 14494. https://doi.org/10.1038/s41598-021-93702-x

[57] J.B. Ulmer, J.J. Donnelly, S.E. Parker, G.H. Rhodes, P.L. Felgner, V.J. Dwarki, S.H. Gromkowski, R.R. Deck, C.M. DeWitt, A. Friedman, L.A. Hawe, K.R. Leander, D. Martinez, H.C. Perry, J.W. Shiver, D.L. Montgomery, M.A. Liu, Heterologous protection against influenza by injection of DNA encoding a viral protein, Science. 259 (1993) 1745-1749. https://doi.org/10.1126/science.8456302

[58] Y. Shi, D.H. Yang, J. Xiong, J. Jia, B. Huang, Y.X. Jin, Inhibition of genes expression of SARS coronavirus by synthetic small interfering RNAs, Cell Research. 15 (2005) 193-200. https://doi.org/10.1038/sj.cr.7290286

[59] Z. Wang, L. Ren, X. Zhao, T. Hung, A. Meng, J. Wang, Y.-G. Chen, Inhibition of Severe Acute Respiratory Syndrome Virus Replication by Small Interfering RNAs in Mammalian Cells, Journal of Virology. 78 (2004) 7523-7527. https://doi.org/10.1128/JVI.78.14.7523-7527.2004

[60] J.R. Petree, K. Yehl, K. Galior, R. Glazier, B. Deal, K. Salaita, Site-Selective RNA Splicing Nanozyme: DNAzyme and RtcB Conjugates on a Gold Nanoparticle, ACS Chemical Biology. 13 (2017) 215-224. https://doi.org/10.1021/acschembio.7b00437

[61] K.C. Petkar, S.M. Patil, S.S. Chavhan, K. Kaneko, K.K. Sawant, NK. Kunda, I.Y. Saleem, An Overview of Nanocarrier-Based Adjuvants for Vaccine Delivery., Pharmaceutics. 13 (2021). https://doi.org/10.3390/pharmaceutics13040455

[62] M.F. Bachmann, G.T. Jennings, Vaccine delivery: A matter of size, geometry, kinetics and molecular patterns, Nature Reviews Immunology. 10 (2010) 787-796. https://doi.org/10.1038/nri2868

[63] M. Henriksen-Lacey, K.S. Korsholm, P. Andersen, Y. Perrie, D. Christensen, Liposomal vaccine delivery systems, Expert Opinion on Drug Delivery. 8 (2011) 505-519. https://doi.org/10.1517/17425247.2011.558081

[64] A.M. Reichmuth, M.A. Oberli, A. Jaklenec, R. Langer, D. Blankschtein, mRNA vaccine delivery using lipid nanoparticles, Therapeutic Delivery. 7 (2016) 319. https://doi.org/10.4155/tde-2016-0006

[65] M.E. Baca-Estrada, M. Foldvari, M. Snider, K. Harding, B. Kournikakis, L.A. Babiuk, P. Griebel, Intranasal immunization with liposome-formulated Yersinia pestis vaccine enhances mucosal immune responses, Vaccine. 18 (2000) 2203-2211. https://doi.org/10.1016/S0264-410X(00)00019-0

[66] R. Pala, V.T. Anju, M. Dyavaiah, S. Busi, S.M. Nauli, Nanoparticle-mediated drug delivery for the treatment of cardiovascular diseases, International Journal of Nanomedicine. 15 (2020) 3741-3769. https://doi.org/10.2147/IJN.S250872

[67] N. Wang, R. Qian, T. Liu, T. Wu, T. Wang, Nanoparticulate carriers used as vaccine adjuvant delivery systems, Critical Reviews in Therapeutic Drug Carrier Systems. 36 (2019) 449-484. https://doi.org/10.1615/CritRevTherDrugCarrierSyst.2019027047

[68] H.Q. Mao, K. Roy, V.L. Troung-Le, K.A. Janes, K.Y. Lin, Y. Wang, J.T. August, K.W. Leong, Chitosan-DNA nanoparticles as gene carriers: Synthesis, characterization and transfection efficiency, Journal of Controlled Release. 70 (2001) 399-421. https://doi.org/10.1016/S0168-3659(00)00361-8

[69] C.R. Alving, R.L. Richards, J. Moss, L.I. Alving, J.D. Clements, T. Shiba, S. Kotani, R.A. Wirtz, W.T. Hockmeyer, Effectiveness of liposomes as potential carriers of vaccines: applications to cholera toxin and human malaria sporozoite antigen, Vaccine. 4 (1986) 166-172. https://doi.org/10.1016/0264-410X(86)90005-8

[70] S. Dhar, W.L. Daniel, D.A. Giljohann, C.A. Mirkin, S.J. Lippard, Polyvalent oligonucleotide gold nanoparticle conjugates as delivery vehicles for platinum(IV) warheads, Journal of the American Chemical Society. 131 (2009) 14652-14653. https://doi.org/10.1021/ja9071282

[71] NanoFlu Influenza Vaccine Phase 3 Study Launches - Precision Vaccinations, (n.d.). https://www.precisionvaccinations.com/novavax-nanoflu-matrix-m-adjuvanted-recombinant-nanoparticle-vaccine-eliciting-antibodies-neutralize (accessed July 18, 2021).

[72] V. F, C. K, F. Vahedifard, K. Chakravarthy, Nanomedicine for COVID-19: the role of nanotechnology in the treatment and diagnosis of COVID-19, Emergent Materials. 4 (2021) 75-99. https://doi.org/10.1007/s42247-021-00168-8

[73] I.D.L. Cavalcanti, M. Cajubá de Britto Lira Nogueira, Pharmaceutical nanotechnology: which products are been designed against COVID-19? 22 (2020) 276. https://doi.org/10.1007/s11051-020-05010-6

[74] G. Chauhan, M.J. Madou, S. Kalra, V. Chopra, D. Ghosh, S.O. Martinez-Chapa, Nanotechnology for COVID-19: Therapeutics and Vaccine Research, ACS Nano. 14 (2020) 7760-7782. https://doi.org/10.1021/acsnano.0c04006

[75] LD. Falo, Advances in skin science enable the development of a COVID-19 vaccine, Journal of the American Academy of Dermatology. 83 (2020) 1226-1227. https://doi.org/10.1016/j.jaad.2020.05.126

[76] S. MD, S. S, C. YH, B. V, C. SK, O.-R. OA, W. DM, C. A, S. M, P. JK, S. NF, M.D. Shin, S. Shukla, Y.H. Chung, V. Beiss, S.K. Chan, O.A. Ortega-Rivera, D.M. Wirth, A. Chen, M. Sack, J.K. Pokorski, N.F. Steinmetz, COVID-19 vaccine development and a potential nanomaterial path forward, Nature Nanotechnology. 15 (2020) 646-655. https://doi.org/10.1038/s41565-020-0737-y

[77] M. Pereira-Silva, G. Chauhan, M.D. Shin, C. Hoskins, M.J. Madou, S.O. Martinez-Chapa, N.F. Steinmetz, F. Veiga, A.C. Paiva-Santos, Unleashing the potential of cell membrane-based nanoparticles for COVID-19 treatment and vaccination, Expert Opinion on Drug Delivery. (2021). https://doi.org/10.1080/17425247.2021.1922387

[78] S.P. Kaur, V. Gupta, COVID-19 Vaccine: A comprehensive status report, Virus Research. 288 (2020). https://doi.org/10.1016/j.virusres.2020.198114

[79] A. Samad, Y. Sultana, M. Aqil, Liposomal Drug Delivery Systems: An Update Review, Current Drug Delivery. 4 (2007) 297-305. https://doi.org/10.2174/156720107782151269

[80] M.J. Copland, M.A. Baird, T. Rades, J.L. McKenzie, B. Becker, F. Reck, P.C. Tyler, NM. Davies, Liposomal delivery of antigen to human dendritic cells, Vaccine. 21 (2003) 883-890. https://doi.org/10.1016/S0264-410X(02)00536-4

[81] G.F.A. Kersten, D.J.A. Crommelin, Liposomes and ISCOMS as vaccine formulations, BBA - Reviews on Biomembranes. 1241 (1995) 117-138. https://doi.org/10.1016/0304-4157(95)00002-9

[82] G.L. Ada, The ideal Vaccine, World Journal of Microbiology & Biotechnology. 7 (1991) 105-109. https://doi.org/10.1007/BF00328978

[83] H. Koike, M. Katsuno, Emerging infectious diseases, vaccines and Guillain-Barré syndrome., Clinical & Experimental Neuroimmunology. (2021). http://www.ncbi.nlm.nih.gov/pubmed/34230841 (accessed July 18, 2021).

[84] A. Stern, H. Markel, The history of vaccines and immunization: familiar patterns, unew challenges, Health Affairs. 24 (2005) 611-621. https://doi.org/10.1377/hlthaff.24.3.611

[85] Centers for Disease Control, Understanding How Vaccines Work, Centers for Disease Control. (2018) 1-2. https://www.cdc.gov/vaccines/hcp/conversations/downloads/vacsafe-understand-color-office.pdf.

[86] A.M. Harandi, D. Medaglini, R.J. Shattock, Vaccine adjuvants: A priority for vaccine research, Vaccine. 28 (2010) 2363-2366. https://doi.org/10.1016/j.vaccine.2009.12.084

[87] I.G. Barr, A. Sjölander, J.C. Cox, ISCOMs and other saponin based adjuvants, Advanced Drug Delivery Reviews. 32 (1998) 247-271 https://doi.org/10.1016/S0169-409X(98)00013-1

[88] M. de Veer, E. Meeusen, New developments in vaccine research--unveiling the secret of vaccine adjuvants., Discovery Medicine. 12 (2011) 195-204.

[89] S.L. Giannini, E. Hanon, P. Moris, M. van Mechelen, S. Morel, F. Dessy, MA Fourneau, B. Colau, J. Suzich, G. Losonksy, M.T. Martin, G. Dubin, M.A. Wettendorff, Enhanced humoral and memory B cellular immunity using HPV16/18 L1 VLP vaccine formulated with the MPL/aluminium salt combination (AS04) compared to aluminium salt only, Vaccine. 24 (2006) 5937-5949. https://doi.org/10.1016/j.vaccine.2006.06.005

[90] Y. Shi, H. HogenEsch, F.E. Regnier, S.L. Hem, Detoxification of endotoxin by aluminum hydroxide adjuvant, Vaccine. 19 (2001) 1747-1752. https://doi.org/10.1016/S0264-410X(00)00394-7

[91] A.S. McKee, M.W. Munks, M.K.L. MacLeod, C.J. Fleenor, N. van Rooijen, J.W. Kappler, P. Marrack, Alum Induces Innate Immune Responses through Macrophage and Mast Cell Sensors, But These Sensors Are Not Required for Alum to Act As an Adjuvant for Specific Immunity, The Journal of Immunology. 183 (2009) 4403-4414. https://doi.org/10.4049/jimmunol.0900164

[92] R.S. Raghuvanshi, Y.K. Katare, K. Lalwani, M.M. Ali, O. Singh, A.K. Panda, Improved immune response from biodegradable polymer particles entrapping tetanus toxoid by use of different immunization protocol and adjuvants, International Journal of Pharmaceutics. 245 (2002) 109-121. https://doi.org/10.1016/S0378-5173(02)00342-3

[93] M. Diwan, M. Tafaghodi, J. Samuel, Enhancement of immune responses by co-delivery of a CpG oligodeoxynucleotide and tetanus toxoid in biodegradable nanospheres, Journal of Controlled Release. 85 (2002) 247-262. https://doi.org/10.1016/S0168-3659(02)00275-4

[94] R.K. Gupta, A.C. Chang, P. Griffin, R. Rivera, G.R. Siber, In vivo distribution of radioactivity in mice after injection of biodegradable polymer microspheres containing 14C-labeled tetanus toxoid, Vaccine. 14 (1996) 1412-1416. https://doi.org/10.1016/S0264-410X(96)00073-4

[95] I. Das, A. Padhi, S. Mukherjee, D.P. Dash, S. Kar, A. Sonawane, Biocompatible chitosan nanoparticles as an efficient delivery vehicle for Mycobacterium tuberculosis lipids to induce potent cytokines and antibody response through activation of γδ T cells in mice, Nanotechnology. 28 (2017). https://doi.org/10.1088/1361-6528/aa60fd

[96] X. Wang, T. Uto, T. Akagi, M. Akashi, M. Baba, Induction of Potent CD8 + T-Cell Responses by Novel Biodegradable Nanoparticles Carrying Human Immunodeficiency Virus Type 1 gp120 , Journal of Virology. 81 (2007) 10009-10016. https://doi.org/10.1128/JVI.00489-07

[97] E. Mohr, A.F. Cunningham, K.M. Toellner, S. Bobat, R.E. Coughlan, R.A. Bird, I.C.M. MacLennan, K. Serre, IFN-γ produced by CD8 T cells induces T-bet-dependent and -independent class switching in B cells in responses to alum-precipitated protein vaccine, Proceedings of the National Academy of Sciences of the United States of America. 107 (2010) 17292-17297. https://doi.org/10.1073/pnas.1004879107

[98] K. Hasegawa, Y. Noguchi, F. Koizumi, A. Uenaka, M. Tanaka, M. Shimono, H. Nakamura, H. Shiku, S. Gnjatic, R. Murphy, Y. Hiramatsu, LJ Old, E. Nakayama, In vitro stimulation of CD8 and CD4 T cells by dendritic cells loaded with a complex of cholesterol-bearing hydrophobized pullulan and NY-ESO-1 protein: Identification of a new HLA-DR15-binding CD4 T-cell epitope, Clinical Cancer Research. 12 (2006) 1921-1927. https://doi.org/10.1158/1078-0432.CCR-05-1900

[99] G. Kumar, T. Ganapathi, C. Revathi, L. Srinivas, V. Bapat, Expression of hepatitis B surface antigen in transgenic banana plants, Planta. 222 (2005) 484-493. https://doi.org/10.1007/s00425-005-1556-y

[100] D.J. Bharali, V. Pradhan, G. Elkin, W. Qi, A. Hutson, S.A. Mousa, Y. Thanavala, Novel nanoparticles for the delivery of recombinant hepatitis B vaccine, Nanomedicine: Nanotechnology, Biology, and Medicine. 4 (2008) 311-317. https://doi.org/10.1016/j.nano.2008.05.006

[101] C. Prego, P. Paolicelli, B. Díaz, S. Vicente, A. Sánchez, Á. González-Fernández, M.J. Alonso, Chitosan-based nanoparticles for improving immunization against hepatitis B infection, Vaccine. 28 (2010) 2607-2614. https://doi.org/10.1016/j.vaccine.2010.01.011

Keyword Index

About the Editors

Dr. Inamuddin is working as Assistant Professor at the Department of Applied Chemistry, Aligarh Muslim University, Aligarh, India. He obtained Master of Science degree in Organic Chemistry from Chaudhary Charan Singh (CCS) University, Meerut, India, in 2002. He received his Master of Philosophy and Doctor of Philosophy degrees in Applied Chemistry from Aligarh Muslim University (AMU), India, in 2004 and 2007, respectively. He has extensive research experience in multidisciplinary fields of Analytical Chemistry, Materials Chemistry, and Electrochemistry and, more specifically, Renewable Energy and Environment. He has worked on different research projects as project fellow and senior research fellow funded by University Grants Commission (UGC), Government of India, and Council of Scientific and Industrial Research (CSIR), Government of India. He has received Fast Track Young Scientist Award from the Department of Science and Technology, India, to work in the area of bending actuators and artificial muscles. He has completed four major research projects sanctioned by University Grant Commission, Department of Science and Technology, Council of Scientific and Industrial Research, and Council of Science and Technology, India. He has published 200 research articles in international journals of repute and nineteen book chapters in knowledge-based book editions published by renowned international publishers. He has published 150 edited books with Springer (U.K.), Elsevier, Nova Science Publishers, Inc. (U.S.A.), CRC Press Taylor & Francis Asia Pacific, Trans Tech Publications Ltd. (Switzerland), IntechOpen Limited (U.K.), Wiley-Scrivener, (U.S.A.) and Materials Research Forum LLC (U.S.A). He is a member of various journals' editorial boards. He is also serving as Associate Editor for journals (Environmental Chemistry Letter, Applied Water Science and Euro-Mediterranean Journal for Environmental Integration, Springer-Nature), Frontiers Section Editor (Current Analytical Chemistry, Bentham Science Publishers), Editorial Board Member (Scientific Reports-Nature), Editor (Eurasian Journal of Analytical Chemistry), and Review Editor (Frontiers in Chemistry, Frontiers, U.K.) He is also guest-editing various special thematic special issues to the journals of Elsevier, Bentham Science Publishers, and John Wiley & Sons, Inc. He has attended as well as chaired sessions in various international and national conferences. He has worked as a Postdoctoral Fellow, leading a research team at the Creative Research Initiative Center for Bio-Artificial Muscle, Hanyang University, South Korea, in the field of renewable energy, especially biofuel cells. He has also worked as a Postdoctoral Fellow at the Center of Research Excellence in Renewable Energy, King Fahd University of Petroleum and Minerals, Saudi Arabia, in the field of polymer electrolyte membrane fuel cells and computational fluid dynamics of polymer electrolyte membrane fuel cells. He is a life member of the Journal of the Indian

Chemical Society. His research interest includes ion exchange materials, a sensor for heavy metal ions, biofuel cells, supercapacitors and bending actuators.

Dr. Tariq Altalhi joined Department of Chemistry at Taif University, Saudi Arabia as Assistant Professor in 2014. He received his doctorate degree from University of Adelaide, Australia in the year 2014 with Dean's Commendation for Doctoral Thesis Excellence. He was promoted to the position of the head of Chemistry Department at Taif university in 2017 and Vice Dean of Science college in 2019 till now. His group is involved in fundamental multidisciplinary research in nanomaterial synthesis and engineering, characterization, and their application in molecular separation, desalination, membrane systems, drug delivery, and biosensing. In 2015, one of his works was nominated for Green Tech awards from Germany, Europ's largest environmental and business prize, amongst top 10 entries.

His interest lies in developing advanced chemistry-based solutions for solid and liquid municipal (both organic and inorganic) waste management. In this direction, he focuses on transformation of solid organic waste to valuable nanomaterials & economic nanostructure. His research work focuses on conversion of plastic bags to carbon nanotubes, fly ash to efficient adsorbent material, etc. Another stream of interests looks at natural extracts and their application in generation of value- added products such as nanomaterials, incense, etc. Through his work as an independent researcher, he has gathered strong management and mentoring skills to run a group of multidisciplinary researchers of various fields including chemistry, materials science, biology, and pharmaceutical science. His publications show that he has developed a wide network of national and international researchers who are leaders in their respective fields. In addition, he has established key contacts with major industries in Kingdom of Saudi Arabia.

Dr. Jorddy Neves Cruz is a researcher at the Federal University of Pará and the Emilio Goeldi Museum. He has experience in multidisciplinary research in the areas of medicinal chemistry, drug design, extraction of bioactive compounds, extraction of essential oils, food chemistry and biological testing. He has published several research articles of international repute. He is currently working as Associate Editor of the Journal of Medicine (Wolters Kluwer, United States).

Dr. Mohammad Luqman has 12+ years of post-PhD experience in Teaching, Research, and Administration. Currently, he is serving as an Assistant Professor of Chemical Engineering in Taibah University, Saudi Arabia. Before joining here, he served as an Assistant Professor in College of Applied Science at A'Sharqiyah University, Oman, and in College of Engineering at King Saud University, Saudi Arabia. He served as a

Research Engineer in SAMSUNG Cheil Industries, South Korea. Moreover, he served as a post-doctoral fellow at Artificial Muscle Research Center, Konkuk University, South Korea, in the field of Ionic Polymer Metal Composites for the development of Artificial Muscles, Robotic Actuators and Dynamic Sensors. He earned his PhD degree in the field of Ionomers (Ion-containing Polymers), from Chosun University, South Korea. He successfully served as an Editor to three books, published by world renowned publishers. He published numerous high-quality papers, and book chapters. He is serving as an Editor and editorial/review board members to many International SCI and Non-SCI journals. He has attracted a few important research grants from industry and academia. His research interests include but not limited to Development of Ionomer/Polyelectrolyte/non-ionic Polymer Nanocomposites/Blends for Smart and Industrial/Engineering Applications.